Charles J.J. Klaver and Wolfgang Böhme
Chamaeleonidae

DAS TIERREICH

The Animal Kingdom

Eine Zusammenstellung und Kennzeichnung der rezenten Tierformen

A Compilation and Characterization of the Recent Animal Groups

Herausgeber/Editors

HEINZ WERMUTH (Wirbeltiere)

MAXIMILIAN FISCHER (Wirbellose)

Teilband **112** Part

Walter de Gruyter · Berlin · New York 1997

Charles J. J. Klaver and Wolfgang Böhme

Chamaeleonidae

Walter de Gruyter · Berlin · New York 1997

The Authors

Drs. Charles J.J. Klaver
University Library Groningen
Postbus 559
NL-9700 AN Groningen
The Netherlands

Prof. Dr. Wolfgang Böhme
Zoologisches Forschungsinstitut und Museum
Alexander Koenig
Adenauerallee 160
D-53113 Bonn
Germany

Scientific and Managing Editor
Dr. Heinz Wermuth
Falkenweg 1
D-71691 Freiberg
Germany

Citation
Das Tierreich, Part 112
Charles J.J. Klaver and Wolfgang Böhme
Verlag Walter de Gruyter & Co., Berlin · New York 1997

♾ printed on acid free paper which falls within the guidelines of the ANSI to ensure permanence and durability.

Library of Congress Cataloging-in-Publication Data

Klaver, Charles,
Chamaeleonidae / Charles J.J. Klaver and Wolfgang Böhme.
p. cm. – (Das Tierreich : Teilbd. 112 = The Animal kingdom . pt. 112)
Includes index
ISBN 3-11-015187-1 (alk. paper)
1. Chameleons. I. Böhme, Wolfgang. II. Title. III. Series; Tierreich ; Teilbd. 112.
QL666.L23K5 1997 97-29347
597.95–dc21 CIP

Die Deutsche Bibliothek – Cataloging-in-Publication Data

Das **Tierreich** : eine Zusammenstellung und Kennzeichnung der rezenten Tierformen = The animal kingdom / Hrsg. Heinz Wermuth ; Maximilian Fischer. – Berlin ; New York : de Gruyter
Teilw. hrsg. von Robert Mertens ; Willi Hennig. – Teilw. hrsg. von F. E. Schulze ... – Teilw. fortges. von R. Hesse

Teilbd. 112. Chamaeleonidae / Charles J.J. Klaver and Wolfgang Böhme. – 1997
ISBN 3-11-015187-1

Typesetting and Printing: Tutte Druckerei GmbH, Salzweg-Passau.
Binding: Lüderitz & Bauer GmbH, Berlin.

Die Liste der rezenten Amphibien und Reptilien *ist*
dem Andenken an

HERRN PROF. DR. ROBERT MERTENS,

Direktor emeritus des Natur-Museums und
Forschungs-Instituts Senckenberg zu Frankfurt am Main,
als Dank der Fachkollegen für seine überragenden
Verdienste um die Herpetologie gewidmet.

Chamaeleonidae

Charles Klaver & Wolfgang Böhme

Drs. Charles J.J. Klaver, University Library Groningen, Postbus 559, 9700 AN Groningen, The Netherlands.
Prof. Dr. Wolfgang Böhme, Zoologisches Forschungsinstitut und Museum Alexander Koenig, Adenauerallee 160, D-53113 Bonn, Germany.

Introduction

The present book contains the third list concerning the Chamaeleonidae in the series of DAS TIERREICH published this century; the two preceding lists were compiled by FRANZ WERNER (1911) and by ROBERT MERTENS (1966). WERNER (1911) recognized 7 (7) species of *Brookesia*, 7 (7) species of *Rhampholeon* and 74 (87) species of *Chamaeleo*, in total 88 (101) species (the numbers between brackets indicate the total number of taxa, i.e. species and subspecies). Fifty-five years later MERTENS (1966) registered 16 (20) species of *Brookesia* (including *Rhampholeon*) and 69 (113) species of *Chamaeleo*, resulting in a total number of 85 (133). In view of the slight decrease in the total number of species recognized and the increase in the number of subspecies, MERTENS expected the total number of chamaeleon taxa to increase little in the future and thought a decrease to be more likely. However, numerous new species, mainly from Madagascar, have been described since 1966, resulting in the present numbers: 23 (24) species of *Brookesia*, 11 (14) species of *Rhampholeon* and 98 (133) species of 'ordinary' chameleons, a total of 132 (171) species. The 'ordinary' chameleons are presently classified in the following genera: *Calumma* with 18 (25) species, *Furcifer* with 19 (20) species, *Bradypodion* with 15 (27) species and *Chamaeleo* with 46 (61) species. The genus *Chamaeleo* is further subdivided into two subgenera, viz. *Chamaeleo* with 14 (24) species and *Trioceros* with 32 (37) species (fide KLAVER & BÖHME 1986). This means that the number of species recognized has increased in 30 years by more than 50%. To find an increase in the same order of magnitude, we have to go back to the end of the last century and compare the total number of species registered by BOULENGER (1887) with that of WERNER (1911), viz. 49 versus 88. This indicates in our opinion that the stasis perceived by MERTENS (1966) was a result of lack of effort in collecting and/or study. The numerous studies of BRYGOO and collaborators during the 1970's, based on recently collected material from Madagascar, proves this by documenting many new and revalidated species. A similar phenomenon for African chameleons is demonstrated by the activities of ARTHUR LOVERIDGE from the 1930's till the 1950's and of numerous German herpetologists who described material collected in German colonial Africa in the second half of the last century and the beginning of the 20th century. Consequently we are inclined to expect the number of known chameleon species to continue to rise in the future as a result of the study of newly collected material from little studied areas, especially mountainous regions, and of the ample museum material of variable 'species' as, e. g. *B. fischeri, B. pumilum* and *Ch. dilepis*. Moreover, even the study of literature may lead to the discovery of new species which we found out ourselves much to our surprise (KLAVER & BÖHME 1988).

In the introduction to his list MERTENS (1966: VI) discussed the status of the various (sub-)genera within the Chamaeleonidae and chose to restrict himself to the use of two generic names, viz.: *Brookesia* and *Chamaeleo*, although he also voiced his expectation that a subdivision of these two genera would occur in the future as a result of further investigations. Subsequently KLAVER & BÖHME (1986) revised the subdivision of the Chamaeleonidae and recognized two subfamilies, viz.: Brookesiinae and Chamaeleoninae. The Brookesiinae were subdivided into two genera, viz.: *Brookesia* and *Rhampholeon*, the Chamaeleoninae into four genera, viz.: *Bradypodion, Calumma, Chamaeleo* and *Furcifer*. The genus *Chamaeleo* was subdivided into two subgenera, viz.: *Chamaeleo* and *Trioceros*. The authors admitted that their classification was provisional, subject to refining in the future as a result of further evidence from additional comparative studies.

Subsequently HOFMAN & al. (1991: 260) found their immunological data to be more or less in concordance with the proposed classification, although they also noticed contradictions between the outcomes of the studies of the different character sets. Nevertheless, this classification was adopted in both the primary literature, e.g. BROADLEY & HOWELL (1991), RAXWORTHY (1991), TILBURY (1991, 1992), EMMRICH (1993, 1994), NEČAS (1994) and RAXWORTHY & NUSSBAUM (1995) and the secondary literature, e.g. GLAW & VENCES (1991) and HENKEL & HEINECKE (1993). Apparently it is considered a better estimate of the phylogenetic relationship of chameleon species than the classification of MERTENS (1966).

Unpublished observations on the hemipenes of *Rhampholeon spectrum* and *R. temporalis* revealed these to possess calyces, whose absence was supposed to be a synapomorphy of the Brookesiinae. Apparently in these two species the plesiomorph calyculate condition is retained. More important, however, is the observation (also unpublished) of an apical ornamentation consisting of four rotulae in the hemipenes of *R. kerstenii*, which falsifies the defining characteristic of the Brookesiinae, viz.: dual apical ornamentation (KLAVER & BÖHME 1986). This problem could be remedied by transferring *R. kerstenii* to the Chamaeleoninae, but this solution would be too convenient and obfuscate taxonomically important observations. Incidentally, these remarkable observations on *R. kerstenii* also falsify the informal subdivision of *Rhampholeon* into two sections as proposed by RIEPPEL (1987). Pending necessary supplementary studies we, therefore, do not retain in this list the subfamilial subdivision we proposed in 1986. We do, however, retain the generic subdivision as we still consider it to reflect the chameleon affinities best. The sequence in which the genera will be presented in this list is not alphabetical, but will conform to the phylogenetic scheme (see KLAVER & BÖHME 1986).

In this list we consider the Chamaeleonidae to include only chameleon species and not agamid species as well (*contra* FROST & ETHERIDGE 1989). We find the argumentation of these authors unconvincing. Their dismissive attitude towards an alternative classification indicates that their parochial approach does not even convince themselves. This is, of course, not surprising as a classification is to be considered a Linnaean convention, i.e. various classifications can be equally consistent with a perceived phylogeny.

We should like to point out that this list is not a purely nomenclatural one, i.e. that we do not accept uncritically the most recent nomenclatural acts in literature. The most recent nomenclatural acts may be, in our opinion, less sufficiently substantiated than earlier ones; in such cases we adopt the older as a guideline for the nomenclature of chameleons in the present list. We are aware that this is a subjective undertaking, but it is in our opinion acceptable when explicitly stated and, wherever necessary, explained in adjoining comments. We find this more realistic than to accept any nomenclatural act indiscriminately or to adhere to what we consider a better substantiated nomenclatural act without comment, but we are aware that readers may be of a different opinion. To illustrate our point of view we refer to MERTENS (1966: VIII), who stated that his list had „... in erster Linie einen nomenklatorischen, nicht taxionomischen Charakter." Unfortunately, he did not always comply with his quoted intention nor with the (then) recent major studies on chameleon taxonomy. For instance, RAND (1963) cogently demonstrated the taxa *schubotzi* and *kinetensis* to be valid species and *altaeelgonis* synonymous with *hoehnelii*, but MERTENS nevertheless chose to treat them as valid subspecies of *rudis*, *bitaeniatus* and *hoehnelii* respectively. Similarly, LOVERIDGE (1957a) and HILLENIUS (1959) argued convincingly that the taxa *vinckei, mertensi* and *marunguensis* are synonymous with *anchietae*, but MERTENS considered them valid subspecies of *anchietae* without argumentation. Whatever the motives of MERTENS might have been, we prefer not to take such an implicit dualistic attitude, but either accept the last nomenclatural act or, if not, explain our different point of view explicitly.

Next to references of original descriptions, new name combinations and presently recognized synonyms, the references used in the two previous Tierreich lists of WERNER (1911) and MERTENS (1966) and in the two comprehensive publications by BRYGOO (1971, 1978) will be cited. As to nomenclature, we should like to point out that we reproduce throughout the list the spelling used in the works referred to without appending a 'sic' in case of an incorrect spelling. This is done to avoid a (necessarily) too prolific use of this indication and we hope that whenever an aberrant spelling occurs, it is the result of a correct representation of the original spelling instead of our own mistake.

Whenever possible information about type specimen(s) is presented and, when appropriate, a short comment will be added to explain relevant issues. It should be noted that collection numbers of type material given in this list may differ from those cited in the literature, because type material may have been largely re-registered (e.g. BMNH material) or transferred to another museum. The literature cited throughout the text is compiled in a separate list of references.

For help in supplying information about type specimens we should like to thank N.B. Ananjeva (ZISP), F. Andreone (MRSN), W.R. Branch (PEM), L. Capocaccia (MCSN), E. Crespo (Univ. de Lisboa), H. Grillitsch (NMW), R. Günther (ZMB), G. Haagner (PEM), H.O. von Hagen (PUM), H. Hohmann (UMB), I. Ineich (MHNP), M. Jacquat (MHNC), B. de Klerk (AM), H.-W. Koepcke (ZMH), E. Kramer (NMB), M. Laudahn (SMF), C. McCarthy (BMNH), A. Resetar (FMNH), G. Rosado (MCZ) and G. Swinney (NMS).

For help finding and for supplying literature we thank F.F.J.M. Pieters and E. Zwart of the Artis Library, Amsterdam and the personnel of the Library of the Teyler's Museum, Haarlem and the Library of the Royal Netherlands Academy of Sciences, Amsterdam.

Collection acronyms

AM – Albany Museum, Grahamstown, South Africa.
AMNH – American Museum of Natural History, New York, U.S.A.
ANSP – Academy of Natural Sciences, Philadelphia, U.S.A.
BMNH – British Museum (Natural History), now The Natural History Museum, London, United Kingdom.
CBMAG – City of Bristol Museum and Art Gallery, Bristol, United Kingdom.
DNSM – Durban Natural Sciences Museum, Durban, South Africa.
FMNH – Field Museum of Natural History, Chicago, U.S.A.
LIV – Liverpool Museum, Liverpool, United Kingdom.
LRC – private collection of L. Raw, Merrivale, South Africa.
MBL – Museu Bocage, Lisbon, Portugal.
MCSN – Museo Civico di Storia Naturali „Giacomo Doria", Genova, Italy.
MCZ – Museum of Comparative Zoology, Cambridge, U.S.A.
MHNC – Musée d'Histoire naturelle, La Chaux-de-Fonds, Switzerland.
MHNG – Muséum d'Histoire Naturelle, Genève, Switzerland.
MHNP – Muséum national d'Histoire Naturelle, Paris, France.
MRAC – Musée Royal de l'Afrique Centrale, Tervuren, Belgium.
MRSN – Museo Regionale di Science Naturali, Torino, Italy.
NMB – Naturhistorisches Museum, Basel, Switzerland.
NMS – National Museums of Scotland, Edinburgh, Scotland.
NMW – Naturhistorisches Museum, Wien, Austria.
NM – Natal Museum, Pietermaritzburg, South Africa.
NRM – Naturhistoriska Riksmuseet, Stockholm, Sweden.
PEM – Port Elizabeth Museum, University of Port Elizabeth, South Africa.
PUM – Philipps-Universität, Marburg, Germany.
SAM – South African Museum, Cape Town, South Africa.
SMF – Senckenberg Museum, Frankfurt a. M., Germany.
SMNS – Staatliches Museum für Naturkunde, Stuttgart, Germany.
TM – Transvaal Museum, Pretoria, South Africa.
UM – Umtali Museum, now Natural History Museum of Zimbabwe (NMZB), Bulawayo, Zimbabwe.
UMB – Übersee-Museum, Bremen, Germany.

UMMZ – University of Michigan, Museum of Zoology, Ann Arbor, U.S.A.
USNM – National Museum of Natural History, Smithsonian Institution, Washington D.C., U.S.A.
ZFMK – Zoologisches Forschungsinstitut und Museum Alexander Koenig, Bonn, Germany.
ZISP – Zoological Institute, St. Petersburg, Russia.
ZMB – Zoologisches Museum, Berlin, Germany.
ZMH – Zoologisches Museum, Hamburg, Germany.
ZSM – Zoologische Staatssammlung, München, Germany.

Systematic Index

If the scientific name in the headline of a paragraph is marked by an asterisk (*), it means that a comment on the taxon concerned is included in the chapter „Appendix" on page 48 and following ones.

Family **Chamaeleonidae***

Genus **Brookesia** GRAY*

1865 *Brookesia* GRAY, Proc. zool. Soc. London, **1864**: 468, 476. – Species typica (by monotypy): *Chamaeleo superciliaris* KUHL 1820.
1933 *Leandria* ANGEL, Bull. Mus. Hist. nat., Paris, (2) **5** (6): 443; figs. 1–2. – Species typica (by monotypy): *Leandria perarmata* ANGEL 1933.
1942 *Evoluticauda* ANGEL, Mém. Acad. malgache, Antananarivo, **36**: 154, 177. – Species typica (by subsequent designation: GUIBÉ 1954: 108, nr. 562; non LOVERIDGE 1956: 2, non LOVERIDGE 1957a: 206): *Brookesia tuberculata* MOCQUARD 1894.
1957 *Leandra* MATTHEY (ex errore), Rev. suisse Zool., **64**: 728.
1988 *Lenadria* BÖHME (ex errore), Bonn. zool. Monogr., **27**: 64.

Distribution: Madagascar and adjacent islets.

Brookesia ambreensis RAXWORTHY & NUSSBAUM

1995 *Brookesia ambreensis* RAXWORTHY & NUSSBAUM, J. Zool., **235**: 534; plate I, fig. 2, 3. – Terra typica: Antomboka River, Montagne d'Ambre National Park (670 m alt.), Antsiranana Fivondronana, Antsiranana (Diégo Suarez) Province, Madagascar. – Holotype (by orig. design.): UMMZ 203635.

Distribution: Mtge. d'Ambre, N. Madagascar.

Brookesia antakarana RAXWORTHY & NUSSBAUM

1995 *Brookesia antakarana* RAXWORTHY & NUSSBAUM, J. Zool., **235**: 537; plate II, figs. 2, 3. – Terra typica: Antomboka River, Montagne d'Ambre National Park (1050 m alt.), Antsiranana Fivondronana, Antsiranana (Diégo Suarez) Province, Madagascar. – Holotype (by orig. design.): UMMZ 200071.

Distribution: Mtge. d'Ambre, N. Madagascar.

Brookesia bekolosy RAXWORTHY & NUSSBAUM

1995 *Brookesia bekolosy* RAXWORTHY & NUSSBAUM, J. Zool., **235**: 540; plate III, figs. 2, 3. – Terra typica: Bekolosy, Manongarivo Special Reserve (1200 m alt.), Ambanja Fivondronana, Antsiranana (Diégo Suarez) Province, Madagascar. – Holotype (by orig. design.): UMMZ 200078.

Distribution: Known only from the type locality.

Brookesia betschi BRYGOO, BLANC & DOMERGUE

1974 *Brookesia betschi* BRYGOO, BLANC & DOMERGUE, Bull. Acad. malgache, Antananarivo, (1973) **51**: 167; figs. 1-3. – Terra typica: Forêt de moyenne altitude (1300 m) du Marojezy, Madagascar. – Holotype (by orig. design.): MHNP 1973.1177.
1978 *Brookesia betschi* – BRYGOO, Faune de Madagascar, **47**: 11; figs. 3-4, map 2.

Distribution: Tsaratanana and Marojezy, Madagascar.

Brookesia bonsi RAMANANTSOA

1980 *Brookesia bonsi* RAMANANTSOA, Bull. Mus. Hist. nat., Paris, (1979) **1** (A): 689; figs. 5-8. – Terra typica: Réserve naturelle no. 8 du Tsingy de Namoroka (Soalala), Madagascar. – Holotype (by orig. design.): MHNP 1978.1992.

Distribution: Known only from the type locality.

Brookesia brygooi RAXWORTHY & NUSSBAUM

1995 *Brookesia brygooi* RAXWORTHY & NUSSBAUM, J. Zool., **235**: 542. – Terra typica: Analavelona, S.W. Madagascar. – Holotype (by orig. design.): MHNP 8219.

Distribution: W. and S.W. Madagascar.

Brookesia decaryi ANGEL*

1939 *Brookesia decaryi* ANGEL, Bull. Mus. Hist. nat., Paris, (1938), **10**: 575. – Terra typica: Côte Ouest, forêt du Massif de l'Ankarafantsika, Madagascar. – Syntypes: MHNP 1938. 158-162 (5 specimens).
1966 *Brookesia decaryi* – MERTENS, Das Tierreich, **83**: 2.
1978 *Brookesia decaryi* – BRYGOO, Faune de Madagascar, **47**: 16; figs. 5-9, map. 3.

Distribution: Ankarafantsika, N.W. Madagascar.

Brookesia dentata MOCQUARD*

1900 *Brookesia dentata* MOCQUARD, Bull. Mus. Hist. nat., Paris, **6**: 345/Bull. Soc. philom., Paris, **2**: 98; plate 2, fig. 4. – Terra typica: Environs de Suberbieville, N.W. Madagascar. – Holotype (by monotypy): MHNP 99.322.
1911 *Brookesia dentata* – WERNER, Das Tierreich, **27**: 43.
1966 *Brookesia dentata* – MERTENS, Das Tierreich, **83**: 2.
1975 *Brookesia ramanantsoai* BRYGOO & DOMERGUE, Bull. Mus. Hist. nat., Paris, (1974) **267**, Zool. **189**: 1773; figs. 3-7. – Terra typica: Forêt d'Ambohiboataba (1300 m alt.), east of Mantasoa, Madagascar. – Holotype (by orig. design.): MHNP 1973. 1401. – Syn.: fide RAXWORTHY & NUSSBAUM 1995: 530.
1978 *Brookesia dentata* – BRYGOO, Faune de Madagascar, **47**: 25; figs. 10-11, map. 4.
1978 *Brookesia ramanantsoai* – BRYGOO, Faune de Madagascar, **47**: 68; figs. 32-33, map 4.

Distribution: Madagascar.

Brookesia ebenaui (BOETTGER)*

1880 *Chamaeleo ebenaui* BOETTGER, Zool. Anz., **3**: 280. – Terra typica: Nossibé (= Nosy Bé), off N.W. Madagascar. – Holotype (by monotypy): SMF 16515.
1887 *Brookesia ebenaui* – BOULENGER, Cat. Liz. Brit. Mus., **3**: 475.
1911 *Brookesia ebenaui* – WERNER, Das Tierreich, **27**: 42.
1966 *Brookesia ebenaui* – MERTENS, Das Tierreich, **83**: 2.

1978 *Brookesia ebenaui* – BRYGOO, Faune de Madagascar, **47**: 28; figs. 12–15, map 3.
1980 *Brookesia legendrei* RAMANANTSOA, Bull. Mus. Hist. nat., Paris, (1979) **1** (A): 685; figs. 1–4. – Terra typica: Réserve naturelle no. 6 de Lokobe (Nosy Bé), Madagascar. – Holotype (by orig. design.): MHNP 1978. 1989. – Syn.: fide RAXWORTHY & NUSSBAUM 1995: 532.

Distribution: Nosy Bé and Madagascar.

Brookesia griveaudi BRYGOO, BLANC & DOMERGUE

1974 *Brookesia griveaudi* BRYGOO, BLANC & DOMERGUE, Bull. Acad. malg., Antananarivo, (1973) **51**: 180; figs. 7–8. – Terra typica: Massif du Marojezy, N.E. Madagascar. – Holotype (by orig. design.): MHNP 1973. 1183.
1978 *Brookesia griveaudi* – BRYGOO, Faune de Madagascar, **47**: 35; figs. 16–17, map 5.

Distribution: N.E. Madagascar.

Brookesia karchei BRYGOO, BLANC & DOMERGUE

1970 *Brookesia karchei* BRYGOO, BLANC & DOMERGUE, Ann. Univ. Madagascar, (Sci) **7**: 271; figs. – Terra typica: Massif du Marojezy (700 m alt.), Madagascar. – Holotype (by orig. design.): MHNP A157.
1978 *Brookesia karchei* – BRYGOO, Faune de Madagascar, **47**: 40; figs. 18–20, map 2.

Distribution: Known only from the type locality.

Brookesia lambertoni BRYGOO & DOMERGUE

1970 *Brookesia lambertoni* BRYGOO & DOMERGUE, Bull. Must. Hist. nat., Paris, (1969) **41**: 1091; figs. 1–2. – Terra typica: Fito, à l'est du pays Sihanaka, Madagascar. – Holotype (by orig. design.): MHNP 21.256.
1978 *Brookesia lambertoni* – BRYGOO, Faune de Madagascar, **47**: 46; figs. 21–22, map 1.

Distribution: Known only from the type locality.

Brookesia lineata RAXWORTHY & NUSSBAUM

1995 *Brookesia lineata* RAXWORTHY & NUSSBAUM, J. Zool., **235**: 543; plate IV, figs. 2 & 3. – Terra typica: Bekolosy, Manongarivo Special Reserve (1200 m alt.), Ambanja Fivondronana, Antsiranana (Diégo Suarez) Province, Madagascar. – Holotype (by orig. design.): UMMZ 200073.

Distribution: Known only from the type locality.

Brookesia lolontany RAXWORTHY & NUSSBAUM

1995 *Brookesia lolontany* RAXWORTHY & NUSSBAUM, J. Zool., **235**: 545; plate V, fig. 3. – Terra typica: Matsabory Lake, Tsaratanana (2050 m alt.), Bealanana Fivondronana, Mahajanga Province, Madagascar. – Holotype (by orig. design.): UMMZ 203629.

Distribution: Tsaratanana, Madagascar.

Brookesia minima BOETTGER*

1893 *Brookesia minima* BOETTGER, Kat. Rept. Mus. Senckenb., **1**: 123, footnote. – Terra typica: Loucoubé auf Nossibé (= Lokoebé, Nosy Bé), off N.W. Madagascar. – Lectotype: SMF 16512 (fide MERTENS, 1967: 64).

1894 *Brookesia tuberculata* MOCQUARD, Bull. Soc. philom., Paris **17**: 3. – Terra typica: Mont d'Ambre, N. Madagascar. – Holotype (by monotypy): MHNP 93.183. – Syn.: fide RAXWORTHY & NUSSBAUM, 1995: 529.
1911 *Brookesia minima* – WERNER, Das Tierreich, **27**: 43.
1911 *Brookesia tuberculata* – WERNER, Das Tierreich, **27**: 43.
1966 *Brookesia minima* – MERTENS, Das Tierreich, **83**: 2.
1966 *Brookesia tuberculata* – MERTENS, Das Tierreich, **83**: 5.
1975 *Brookesia peyrierasi* BRYGOO & DOMERGUE, Bull. Mus. Hist. nat., Paris, (1974) **267** (Zool. **189**): 1779. – Terra typica: Nosy Mangabe, Antongil Bay, Madagascar. – Holotype (by orig. design.): MHNP 1968.190. – Syn.: fide RAXWORTHY & NUSSBAUM 1995: 529.
1978 *Brookesia minima* – BRYGOO, Faune de Madagascar, **47**: 49; fig. 23, map. 4.
1978 *Brookesia peyrierasi* – BRYGOO, Faune de Madagascar, **47**: 66; fig. 31, map 4.
1978 *Brookesia tuberculata* – BRYGOO, Faune de Madagascar, **47**: 97; figs. 47–49, map 4.

Distribution: Nosy Mangabe, Nosy Bé and N. Madagascar.

Brookesia nasus BOULENGER

1887 *Brookesia nasus* BOULENGER, Cat. Liz. Brit. Mus., **3**: 475; plate XL, fig. 4. – Terra typica: Ekongo (= Ikongo?), Madagascar. – Holotype (by monotypy): BMNH 1879.10.21.8
1978 *Brookesia nasus* – BRYGOO, Faune de Madagascar, **47**: 53.

Distribution: S.E. Madagascar.

Brookesia nasus nasus BOULENGER

1911 *Brookesia nasus* – WERNER, Das Tierreich, **27**: 44.
1927 *Brookesia betsileana* AHL, Zool. Anz., **72**: 48. – Terra typica: Betsileo, Madagascar. – Holotype (by monotypy): ZMB 30564. – Syn.: fide ANGEL 1942: 178.
1966 *Brookesia nasus* – MERTENS, Das Tierreich, **83**: 3.
1972 *Brookesia nasus nasus* – BRYGOO, BLANC & DOMERGUE, Bull. Mus. Hist. nat., Paris, **56** (Zool. **42**): 595; figs. 1–2.
1978 *Brookesia nasus nasus* – BRYGOO, Faune de Madagascar, **47**: 53; figs. 25–28, map 2.

Distribution: S.E. Madagascar.

Brookesia nasus pauliani BRYGOO, BLANC & DOMERGUE

1972 *Brookesia nasus pauliani* BRYGOO, BLANC & DOMERGUE, Bull. Mus. Hist. nat., Paris **56**, (zool., **42**): 595; figs. 3–4. – Terra typica: Manjarivolo, l'Andringitra (1620–1650 m alt.), Madagascar. – Holotype (by orig. design.): MHNP 1971.276.
1978 *Brookesia nasus pauliani* – BRYGOO, Faune de Madagascar, **47**: 61; fig. 29, map 2.

Distribution: Known only from the type locality.

Brookesia perarmata (ANGEL)

1933 *Leandria perarmata* ANGEL, Bull. Mus. Hist. nat., Paris **5**: 444; figs. 1–2. – Terra typica: Antsingy (alt. 300 m), province du Ménabé, Madagascar. – Holotype (by monotypy): MHNP 33.145.
1966 *Brookesia perarmata* – MERTENS, Das Tierreich, **83**: 3.
1978 *Brookesia perarmata* – BRYGOO, Faune de Madagascar, **47**: 63; fig. 30, map 2.

Distribution: Known only from the type locality.

Brookesia stumpffi BOETTGER*

1894 *Brookesia stumpffi* BOETTGER, Zool. Anz., **17**: 182. – Terra typica restricta (hoc loco): Insel Nossibé (= Nosy Bé), off N.W. Madagascar. – Lectotype (hoc loco): SMF 16475.
1911 *Brookesia stumpffi* – WERNER, Das Tierreich, **27**: 42; figs. 15–16.
1966 *Brookesia stumpffi* – MERTENS, Das Tierreich, **83**: 4.
1978 *Brookesia stumpffi* – BRYGOO, Faune de Madagascar, **47**: 71; figs. 34–38, map 5.

Distribution: Nosy Bé and N. Madagascar.

Brookesia superciliaris (KUHL)*

1820 *Chamaeleo superciliaris* KUHL, Beitr. Zool. vergl. Anat., **1**: 103. – Terra typica: not given. – Type(s): BMNH; not registered as type(s).
1827 *Chameleo brookesiana* GRAY, Philos. Mag., London, **2**: 409. – Terra typica: probably from Africa. – Holotype (by monotypy): BMNH 1946.8.21.80. – Syn.: fide BOULENGER 1887: 474.
1831 *Chameleo brookesii* GRAY (nomen substitutum pro *Chameleo brookesiana* GRAY 1827), in GRIFFITH, Anim. Kingd., **9** (Synops): 53.
1843 *Chamaeleon (Furcifer) brookesii* FITZINGER (nomen substitutum pro *Chameleo brookesiana* GRAY 1827), Syst. Rept.: 42.
1845 *Chameleo brooksianus* GRAY (ex errore), Cat. Liz. Brit. Mus.: 270.
1865 *Brookesia superciliaris* – GRAY, Proc. zool. Soc. London, **1864**: 477.
1885 *Chamaeleo superciliosus* MÜLLER (ex errore), Verh. naturf. Ges. Basel, **7**: 164.
1911 *Brookesia superciliaris* – WERNER, Das Tierreich, **27**: 41.
1942 *Chamaeleon brookesia* ANGEL (ex errore pro *Chameleo brookesiana* GRAY 1827), Mém. Acad. malg., Antananarivo, **36**: 180.
1966 *Brookesia superciliaris* – MERTENS, Das Tierreich, **83**: 4.
1978 *Brookesia superciliaris* – BRYGOO, Faune de Madagascar, **47**: 80; figs. 39–41, map 6.

Distribution: E. Madagascar.

Brookesia therezieni BRYGOO & DOMERGUE

1970 *Brookesia therezieni* BRYGOO & DOMERGUE, Bull. Mus. Hist. nat., Paris, (1969), **4**: 1093; figs. 3–5. – Terra typica: Périnet (= Andasibe), sous-préfecture de Moramanga, Madagascar. – Holotype (by orig. design.): MHNP A163.
1978 *Brookesia therezieni* – BRYGOO, Faune de Madagascar. **47**: 88; figs. 42–43, map 6.

Distribution: E. Madagascar.

Brookesia thieli BRYGOO & DOMERGUE

1969 *Brookesia thieli* BRYGOO & DOMERGUE, Bull. Mus. Hist. nat., Paris, (1968), **40**: 1103; figs. 1–4. – Terra typica: Ambodimanga, canton de Beparasy, sous-préfecture de Moramanga, Madagascar. – Holotype (by orig. design.): MHNP 1968.132.
1971 *Brookesia antoetrae* BRYGOO & DOMERGUE, Bull. Mus. Hist. nat., Paris, (1970) **42**: 830; figs. 1–3. – Terra typica: Ambahona, Antoetra, sous-préfecture d'Ambositra, Madagascar. – Holotype (by orig. design.): MHNP 1969.112. – Syn.: fide RAXWORTHY & NUSSBAUM 1995: 530.
1978 *Brookesia antoetrae* – BRYGOO, Faune de Madagascar, **47**: 7; figs. 1–2, map 1.
1978 *Brookesia thieli* – BRYGOO, Faune de Madagascar, **47**: 92; figs. 44–47, map 1.

Distribution: N. & E. Madagascar.

Brookesia vadoni BRYGOO & DOMERGUE

1968 *Brookesia vadoni* BRYGOO & DOMERGUE, Bull. Mus. Hist. nat., Paris, **40**: 682; figs. 1–4. – Terra typica: Vallée de la rivière Iaraka, near Masoala (600–1000 m alt.), Madagascar. – Holotype (by orig. design.): MHNP 1968.3.

1978 *Brookesia vadoni* – BRYGOO, Faune de Madagascar, **47**: 101; figs. 50–51, map 2.

Distribution: Known only from the type locality.

Brookesia valerieae RAXWORTHY

1991 *Brookesia valerieae* RAXWORTHY, J. Zool., London, **224**: 17; plate V-Vb. – Terra typica: Manongarivo Special Reserve (350 m alt.), N.W. Madagascar. – Holotype (by orig. design.): BMNH 1988.8.

Distribution: Manongarivo Special Reserve, Madagascar.

Genus Rhampholeon GÜNTHER

1874 *Rhampholeon* GÜNTHER, Proc. zool. Soc. London, **1874**: 443, 444; fig., plate 57. – Species typica (by monotypy): *Chamaeleo spectrum* BUCHHOLZ 1874.
1874 *Rampholeon* GÜNTHER (ex errore), Proc. zool. Soc. London, **1874**: 443.
1893 *Rhampholeo* PFEFFER (ex errore), Mitt. naturhist. Mus., Hamburg, (1892), **10**: 76.
1906 *Rhampholeo* KLAPTOCZ (ex errore), Zool. Jb., Anat., **23**: 188.
1906 *Rampholeo* KLAPTOCZ (ex errore), Zool. Jb., Anat., **23**: 188.
1937 *Rhompholeon* PASQUAL (ex errore), Niger. Fld., **6**: 33.
1956 *Bicuspis* LOVERIDGE, Breviora, Mus. comp. Zool., **59**: 2. – Species typica (by orig. design.): *Rhampholeon marshalli* BOULENGER 1906.

Distribution: Tropical Africa.

Rhampholeon boulengeri STEINDACHNER*

1911 *Rhampholeon boulengeri* STEINDACHNER, Anz. Akad. Wiss., Wien, math.-naturw. Kl., **48**: 178. – Terra typica: Nordwestlichen Ufers des Tanganyikasees, Kongo (= Zaire). – Syntypes: NMW 16000: 1–3 (3 specimens).
1911 *Rhampholeon boulengeri* – WERNER, Das Tierreich, **27**: 46.
1942 *Brookesia spectrum boulengeri* – LOVERIDGE, Bull. Mus. comp. Zool., **91**: 240, 370.
1955 *Rhampholeon spectrum goulengeri* DE WITTE (ex errore), Bull. Acad. roy. Sci. colon., **1**: 209.
1966 *Brookesia spectrum boulengeri* – MERTENS, Das Tierreich, **83**: 4.

Distribution: E. Zaire, Rwanda, Burundi, Uganda, Kenya and Tanzania.

Rhampholeon brachyurus GÜNTHER

1893 *Rhampholeon brachyurus* GÜNTHER, Proc. zool. Soc. London, **1892**: 557; plate 34, figs. 2–2a. – Terra typica: Shiré Highlands, south of Lake Nyassa, principally upon Mount Zomba and Mount Milanje, Nyassaland (= Malawi). – Syntypes: BMNH 1946.8.70–71 (2 specimens).
1911 *Rhampholeon brevicaudatus* – WERNER (partim), Das Tierreich, **27**: 46.
1951 *Brookesia ionidesi* LOVERIDGE, Bull. Mus. comp. Zool., **106**: 179. – Terra typica: Kilwa, southern province, Tanganyika Territory (= Tanzania). – Holotype (by orig. design.): MCZ 51301. – Syn.: fide BROADLEY in TILBURY 1992: 7, and BROADLEY & HOWELL 1991: 14.
1951 *Brookesia brachyura* – LOVERIDGE, Bull. Mus. comp. Zool., **106**: 182.
1953 *Brookesia brachyura brachyura* – LOVERIDGE, Bull. Mus. comp. Zool., **110**: 192.
1953 *Brookesia brachyura ionidesi* – LOVERIDGE, Bull. Mus. comp. Zool., **110**: 193.
1966 *Brookesia brachyura* – MERTENS, Das Tierreich, **83**: 1.
1966 *Brookesia brachyura brachyura* – MERTENS, Das Tierreich, **83**: 1.
1966 *Brookesia brachyura ionidesi* – MERTENS, Das Tierreich, **83**: 1.

1986 *Rhampholeon brachyurus* – KLAVER & BÖHME, Bonn. zool. Monogr., **22**: 58.

Distribution: S.E. Tanzania, Malawi, and Mozambique.

Rhampholeon brevicaudatus (MATSCHIE)

1892 *Chamaeleon (Brookesia) brevicaudatus* MATSCHIE, S.Ber. Ges. naturf. Fr. Berlin, **1892**: 107. – Terra typica: Derema, Usambara Mts., Tanganyika (= Tanzania). – Holotype (by monotypy): ZMB 11349.

1893 *Rhampholeo boettgeri* PFEFFER, Mitt. naturhist. Mus., Hamburg, (1892) **10**: 76; plate 1, figs. 6–7. – Terra typica: Bagamoyo, Tanganyika. – Type(s): ZMH; presumed lost by fire in 1943 (H.-W. Koepcke, in litt. 21.VII.1994). – Syn.: fide WERNER 1902: 431.

1911 *Rhampholeon brevicaudatus* – WERNER (partim) Das Tierreich, **27**: 46.

1942 *Brookesia brevicaudata* – LOVERIDGE, Bull. Mus. comp. Zool., **91**: 371.

1966 *Brookesia brevicaudata* – MERTENS, Das Tierreich, **83**: 1.

1986 *Rhampholeon brevicaudatus* – KLAVER & BÖHME, Bonn. zool. Monogr., **22**: 58.

Distribution: E. Tanzania.

Rhampholeon chapmanorum TILBURY*

1992 *Rhampholeon chapmani* TILBURY, Trop. Zool., **5**: 2; figs. 1–2. – Terra typica: above Chididi Police post (= Mbeu Police post (940 m alt.), Malawi Hill, Natundu Hills Range, southern Malawi. – Holotype (by orig. design.): BMNH 1988.399.

Distribution: Known only from the type locality.

Rhampholeon kerstenii (PETERS)

1868 *Chamaeleo kerstenii* PETERS, M.Ber. Akad. Wiss., Berlin, **1868**: 449. – Terra typica: Wanga, bei Mombas (= Mombasa), Kenya. – Holotype (by monotypy): ZMB 5516.

1880 *Rhampholeon kerstenii* – GÜNTHER, Ann. Mag. nat. His. London, **6**: 238.

1936 *Brookesia kerstenii* – LOVERIDGE, Bull. Mus. comp. Zool., **79**: 335.

1966 *Brookesia kerstenii* – MERTENS, Das Tierreich, **83**: 2.

1986 *Rhampholeon kerstenii* – KLAVER & BÖHME, Bonn. zool. Monogr., **22**: 58.

Distribution: Kenya, Somalia and Tanzania.

Rhampholeon kerstenii kerstenii (PETERS)

1911 *Rhampholeon kerstenii* – WERNER, Das Tierreich, **27**: 47.

1936 *Brookesia kerstenii kerstenii* – LOVERDIGE, Bull. Mus. comp. Zool., **79**: 335.

1966 *Brookesia kerstenii kerstenii* – MERTENS, Das Tierreich, **83**: 2.

Distribution: Kenya and Tanzania.

Rhampholeon kerstenii robecchii BOULENGER

1892 *Rhampholeon robecchii* BOULENGER, Ann. Mus. civ. Stor. nat., Genova, **32**: 13; plate 1, figs. 3–3a. – Terra typica: Wuorandi (= Warandi), near Obbia, Somalia. – Holotype (by orig. design.): MCSN-EE 29079; presumed lost in the floods of 1970 (L. Capocaccia, pers.comm.).

1897 *Rhampholeon mandera* Meek, Field Mus. Nat. Hist., (Zool.) **1**: 183. – Terra typica: Mandera, British Somaliland. – Holotype (by orig. design.): FMNH 370. – Syn.: fide: WERNER 1902: 430).

1911 *Rhampholeon robecchii* – WERNER, Das Tierreich, **27**: 46.

1942 *Rhampholeon kerstenii robecchii* – PARKER, Bull. Mus. comp. Zool., **91**: 80.

1966 *Brookesia kerstenii robecchii* – MERTENS, Das Tierreich, **83**: 2.

Distribution: Somalia and N. Kenya.

Rhampholeon marshalli BOULENGER

1906 *Rhampholeon marshalli* BOULENGER, Ann. Mag. nat. Hist., London, **18**: 346; fig. – Terra typica: Chirinda forest (4500 ft. alt.), S.E. Mashonaland, Zimbabwe. – Holotype (by monotypy): BMNH 1946.8.21.72.
1951 *Chamaeleo marshalli* – LOVERIDGE, Bull. Mus. comp. Zool., **106**: 182; footnote.
1971 *Rhampholeon marshalli* – BROADLEY, Arnoldia, **5**: 1; plate 1.

Distribution: E. Zimbabwe and W. Mozambique.

Rhampholeon marshalli marshalli BOULENGER

1911 *Rhampholeon marshalli* – WERNER, Das Tierreich, **27**: 47.
1966 *Chamaeleo marshalli* – MERTENS, Das Tierreich, **83**: 19.
1971 *Rhampholeon marshalli marshalli* – BROADLEY, Arnoldia, **5**: 1; plate 1.

Distribution: E. Zimbabwe.

Rhampholeon marshalli gorongosae BROADLEY

1971 *Rhampholeon marshalli gorongosae* BROADLEY, Arnoldia, **5**: 2; plate 2. – Terra typica: Gorongoso Mts. (1200 m alt.), Manica e Sofale district, Mozambique. – Holotype (by orig. design.): UM 21786.

Distribution: Known only from the type locality.

Rhampholeon nchisiensis (LOVERIDGE)

1953 *Brookesia nchisiensis* LOVERIDGE, Bull. Mus. comp. Zool., **110**: 190; plate 3, fig. 1. – Terra typica: Nchisi forest (5000 m alt.), Nchisi Mt., Nyassaland (= Malawi). – Holotype (by orig. design.): MCZ 50681.
1966 *Brookesia nchisiensis* – MERTENS, Das Tierreich, **83**: 3.
1986 *Rhampholeon nchisiensis* – KLAVER & BÖHME, Bonn. zool. Monogr., **22**: 58.

Distribution: Malawi and S. Tanzania.

Rhampholeon platyceps GÜNTHER

1893 *Rhampholeon platyceps* GÜNTHER, Proc. zool. Soc. London, **1892**: 556; plate 34, figs. 1–1a. – Terra typica: Shire Highlands, south of Lake Nyassa, principally upon Mount Zomba and Mount Milanje, Nyassaland (= Malawi). – Terra typica restricta (fide STEVENS 1974: 13): Mulanje Mountain, Malawi. – Holotype (by monotypy): BMNH 1946.8.21.73.
1896 *Rampholeon platiceps* BOCAGE (ex errore), J. Acad. Sci. Lisboa; **4**: 103.
1933 *Brookesia platyceps* – LOVERIDGE, Bull. Mus. comp. Zool., **74**: 343.
1966 *Brookesia platyceps* – MERTENS, Das Tierreich, **83**: 3.
1966 *Rhampholeon platiceps* MERTENS (ex errore), Das Tierreich, **83**: 36.
1986 *Rhampholeon platyceps* – KLAVER & BÖHME, Bonn. zool. Monogr., **22**: 58.

Distribution: Malawi.

Rhampholeon platyceps platyceps GÜNTHER

1911 *Rhampholeon platyceps* – WERNER, Das Tierreich, **27**: 47.

1953 *Brookesia platyceps platyceps* – LOVERIDGE, Bull. Mus. comp. Zool., **110**: 193.
1966 *Brookesia platyceps platyceps* – MERTENS, Das Tierreich, **83**: 3.

Distribution: Malawi.

Rhampholeon platyceps carri (LOVERIDGE)

1953 *Brookesia platyceps carri* LOVERIDGE, Bull. Mus. comp. Zool., **110**: 194; plate 3, fig. 2a. – Terra typica: Lichenya plateau (6000 ft.), Mlanje Mountain, Nyassaland (= Malawi). – Holotype (by orig. design.): AMNH 67823.
1966 *Brookesia platyceps carri* – MERTENS, Das Tierreich, **83**: 3.

Distribution: Known only from the type locality.

Rhampholeon spectrum (BUCHHOLZ)*

1874 *Chamaeleo spectrum* BUCHHOLZ, M.Ber. Akad. Wiss. Berlin, **1874**: 298, footnote; plate, figs. 5–6. – Terra typica: Victoria and Bonjongo, Cameroon – Terra typica restricta: Bonjongo, Cameroon (fide MERTENS 1938: 43). – Syntypes: ZMB 8026 (2 specimens).
1874 *Rhampholeon spectrum* – GÜNTHER, Proc. zool. Soc. London, **1874**: 443; fig.
1911 *Rhampholeon affinis* STEINDACHNER, Anz. Akad. Wiss. Wien, math.-naturw. Kl., **48**: 178. – Terra typica: Beni, Kivudistr., Kongo (= Zaire). – Syntypes: NMW 16001: 1–2 (2 specimens). – Syn.: hoc loco.
1911 *Rhampholeon spectrum* – WERNER, Das Tierreich, **27**: 45; figs. 17–20.
1951 *Brookesia spectrum spectrum* – LOVERIDGE, Bull. Mus. comp. Zool., **106**: 182.
1966 *Brookesia spectrum* – MERTENS, Das Tierreich, **83**: 3.
1966 *Brookesia spectrum spectrum* – MERTENS, Das Tierreich, **83**: 3.
1973 *Rhampholeon spectrum* – WRIGHT & BROADLEY, Bull. South. California, Acad. Sci., **72**: 165.

Distribution: Bioko (Fernando Po), Cameroon, Gabon, Congo, Zaire.

Rhampholeon temporalis (MATSCHIE)

1892 *Chamaeleon (Brookesia) temporalis* MATSCHIE, S.Ber. Ges. naturf. Fr. Berlin, **1892**: 108. – Terra typica: Derema, Usambara Mts., Tanganyika (= Tanzania). – Holotype (by monotypy): ZMB 11351.
1911 *Chamaeleon temporalis* – WERNER, Das Tierreich, **27**: 33.
1933 *Brookesia temporalis* – LOVERIDGE, Bull. Mus. comp. Zool., **74**: 342.
1966 *Brookesia temporalis* – MERTENS, Das Tierreich, **83**: 4.
1986 *Rhampholeon temporalis* – KLAVER & BÖHME, Bonn. zool. Monogr., **22**: 58.

Distribution: Usambara Mts., Tanzania.

Genus **Calumma** GRAY*

1865 *Calumma* GRAY, Proc. zool. Soc. London, **1864**: 468, 476; fig. – Species typica (by monotypy): *Chameleon cucullatus* GRAY 1831.
1865 *Crassonota* GRAY, Proc. zool. Soc. London, **1864**: 468, 477. – Species typica (by monotypy): *Chameleo nasutus* GRAY (= DUMÉRIL & BIBRON 1836).
1865 *Archaius* GRAY, Proc. zool. Soc. London, **1864**: 475. – Species typica (by subsequent designation: LOVERIDGE 1957a: 197): *Chamaeleo tigris* KUHL 1820.
1873 *Calumna* MASHALL (ex errore), Nomencl. zool.: 45.
1942 *Columna* ANGEL (ex errore), Mém. Acad. malg., Antananarivo, **36**: 175.

Distribution: Madagascar, adjacent islets and Seychelles.

Calumma boettgeri (BOULENGER)

1888 *Chamaeleon boettgeri* BOULENGER, Ann. Mag. nat. Hist., London, **1**: 23; plate 2, fig. 3. – Terra typica: Nossi Bé (= Nosy Bé), off N.W. Madagascar. – Syntypes: BMNH 1946.8.21.56–58 (3 specimens).
1903 *Chamaeleo macrorhinus* BARBOUR, Proc. biol. Soc. Washington, **16**: 62. – Terra typica: Madagascar. – Holotype (by orig. design.): MCZ 5988. – Syn.: fide MOCQUARD 1909: 6.
1911 *Chamaeleon boettgeri* – WERNER, Das Tierreich, **27**: 39; fig 1 b.
1921 *Chamaeleon boettgerti* ANGEL (ex errore), Bull. Mus. Hist. nat. Paris, **27**: 411.
1966 *Chamaeleo boettgeri* – MERTENS, Das Tierreich, **83**: 8.
1971 *Chamaeleo boettgeri* – BRYGOO, Faune de Madagascar, **33**: 76; figs. 15–16, map 9.
1978 *Chamaeleo boettgeri* – BRYGOO, Faune de Madagascar, **47**: 118; fig. 56.
1986 *Calumma boettgeri* – KLAVER & BÖHME, Bonn. zool. Monogr., **22**: 58.

Distribution: Nosy Bé and N. Madagascar.

Calumma brevicornis (GÜNTHER)

1879 *Chamaeleon brevicornis* GÜNTHER, Proc. zool. Soc. London, **1879**: 148; plate 12, fig. A. – Terra typica: Near Antananarivo, Madagascar. – Holotype (by monotypy): BMNH 1946.8.21.41.
1971 *Chamaeleo brevicornis* – BRYGOO, Faune de Madagascar, **33**: 82.
1978 *Chamaeleo brevicornis* – BRYGOO, Faune de Madagascar, **47**: 172.
1986 *Calumma brevicornis* – KLAVER & BÖHME, Bonn. zool. Monogr., **22**: 58.

Distribution: Nosy Boraha and N. and E. Madagascar.

Calumma brevicornis brevicornis (GÜNTHER)

1879 *Chamaeleon gularis* GÜNTHER, Proc. zool. Soc. London, **1879**: 149; plate 12, fig. B. – Terra typica: Near Antananarivo, Madagascar. – Holotype (by monotypy): BMNH 1946.8.21.50. – Syn.: fide GÜNTHER 1882: 263.
1911 *Chamaeleon brevicornis* – WERNER, Das Tierreich, **27**: 31; fig. 9.
1966 *Chamaeleo brevicornis* – MERTENS, Das Tierreich, **83**: 9.
1970 *Chamaeleo brevicornis brevicornis* – BRYGOO & DOMERGUE, Bull. Must. Hist. nat. Paris, **42**: 319; fig. 4.
1971 *Chamaeleo brevicornis brevicornis* – BRYGOO, Faune de Madagascar, **33**: 82; fig. 17–21, map. 10.
1978 *Chamaeleo brevicornis brevicornis* – BRYGOO, Faune de Madagascar, **47**: 119.

Distribution: Nosy Boraha and N. and E. Madagascar.

Calumma brevicornis hilleniusi (BRYGOO, BLANC & DOMERGUE)*

1973 *Chamaeleo brevicornis hilleniusi* BRYGOO, BLANC & DOMERGUE, Bull. Soc. zool. France, **98**: 120; figs. 1–4. – Terra typica: Station de Manjakatompa, l'Ankaratra, Madagascar. – Holotype (by. orig. design.): MHNP 1972.48.
1978 *Chamaeleo brevicornis hilleniusi* – BRYGOO, Faune de Madagascar, **47**: 120; figs. 57–60.

Distribution: Known only from the type locality.

Calumma brevicornis tsarafidyi (BRYGOO & DOMERGUE)*

1970 *Chamaeleo brevicornis tsarafidyi* BRYGOO & DOMERGUE, Bull. Mus. Hist. nat. Paris, **42**: 319; figs. 1–3. – Terra typica: Forêt de Tsarafidyi, sous-préfecture d'Ambohimahasoa, Madagascar. – Holotype (by orig. design.): MHNP A172.

1971 *Chamaeleo brevicornis tsarafidyi* – Brygoo, Faune de Madagascar, **33**: 95; fig. 22, map. 10.
1978 *Chamaeleo brevicornis tsarafidyi* – Brygoo, Faune de Madagascar, **47**: 172.

Distribution: Known only from the type locality.

Calumma capuroni (Brygoo, Blanc & Domergue)

1972 *Chamaeleo capuroni* Brygoo, Blanc & Domergue, Bull. Mus. Hist. nat. Paris, **56** (Zool., **42**): 601; figs. 1–4. – Terra typica: Massif situé à l'extremité septentrionale des chaînes Anosyennes (1900–1950 m alt.), Madagascar. – Holotype (by orig. design.): MHNP A332.
1978 *Chamaeleo capuroni* – Brygoo, Faune de Madagascar, **47**: 126; figs. 62–64.
1986 *Calumma capuroni* – Klaver & Böhme, Bonn. zool. Monogr., **22**: 58.

Distribution: Known only from the type locality.

Calumma cucullata (Gray)

1831 *Chameleon cucullatus* Gray in Griffith, Animal Kingdom, **9** (Synops.): 54. – Terra typica: Madagascar. – Holotype (by monotypy): BMNH 1946.8.21.82.
1911 *Chamaeleon cucullatus* – Werner, Das Tierreich, **27**: 32.
1966 *Chamaeleo cucullatus* – Mertens, Das Tierreich, **83**: 12.
1971 *Chamaeleo cucullatus* – Brygoo, Faune de Madagascar, **33**: 103; figs. 25–27, map 12.
1978 *Chamaeleo cucullatus* – Brygoo, Faune de Madagascar, **47**: 131; fig. 65.
1986 *Calumma cucullata* – Klaver & Böhme, Bonn. zool. Monogr., **22**: 58.

Distribution: N.E. Madagascar.

Calumma fallax (Mocquard)

1900 *Chamaeleon fallax* Mocquard, Bull. Mus. Hist. nat. Paris, **6**: 345/Bull. Soc. philom., Paris, **2**: 96; plate 2, fig. 2. – Terra typica: Forêt d'Ikongo, Madagascar. – Syntypes: MHNP 99.317–318, 88.24 and 90.430–432 (6 specimens).
1911 *Chamaeleon fallax* – Werner, Das Tierreich, **27**: 40.
1966 *Chamaeleo fallax* – Mertens, Das Tierreich, **83**: 13.
1971 *Chamaeleo fallax* – Brygoo, Faune de Madagascar, **33**: 113; figs. 31–32, map 9.
1978 *Chamaeleo fallax* – Brygoo, Faune de Madagascar, **47**: 132.
1986 *Calumma fallax* – Klaver & Böhme, Bonn. zool. Monogr., **22**: 58.

Distribution: E. Madagascar.

Calumma furcifer (Vaillant & Grandidier)

1880 *Chamaeleo furcifer* Vaillant & Grandidier, Bull. Soc. philom., Paris, **4**: 148. – Terra typica: Côte Est de Madagascar. – Holotype (by monotypy): MHNP 5539.
1911 *Chamaeleon furcifer* – Werner, Das Tierreich, **27**: 30.
1913 *Chamaeleon chauvini* Methuen & Hewitt, Ann. Transvaal Mus., Pretoria, **3**: 191; plate 11, fig. 4. – Terra typica: Tamatave, Madagascar. – Holotype (by monotypy): TM 4237. – Syn.: fide Angel 1942: 171.
1966 *Chamaeleo furcifer* – Mertens, Das Tierreich, **83**: 15.
1971 *Chamaeleo furcifer* – Brygoo, Faune de Madagascar, **33**: 120; fig. 33A-B, map 13.
1978 *Chamaeleo furcifer* – Brygoo, Faune de Madagascar, **47**: 172.
1986 *Calumma furcifer* – Klaver & Böhme, Bonn. zool. Monogr., **22**: 58.

Distribution: E. Madagascar.

Calumma gallus (GÜNTHER)

1877 *Chamaeleon gallus* GÜNTHER, Ann. Mag. nat. Hist., London, **19**: 315; plate 16, fig. B. – Terra typica: Mahanoro, S. of Tamatave, Madagascar. – Holotype (by monotypy): BMNH 1946.8.21.55.
1911 *Chamaeleon gallus* – WERNER, Das Tierreich, **27**: 40.
1966 *Chamaeleo gallus* – MERTENS, Das Tierreich, **83**: 15.
1971 *Chamaeleo gallus* – BRYGOO, Faune de Madagascar, **33**: 124; figs. 34–36, map 9.
1978 *Chamaeleo gallus* – BRYGOO, Faune de Madagascar, **47**: 132.
1986 *Calumma gallus* – KLAVER & BÖHME, Bonn. zool. Monogr., **22**: 58.

Distribution: Île aux Prunes, E. Madagascar and ? Nosy Bé.

Calumma gastrotaenia (BOULENGER)*

1888 *Chamaeleon gastrotaenia* BOULENGER, Ann. Mag. nat. Hist., London, **1**: 103; plate 5, fig. 2. – Terra typica: Madagascar. – Syntypes: BMNH 1946.8.18.60–61 (2 specimens).
1971 *Chamaeleo gastrotaenia* BRYGOO, Faune de Madagascar, **33**: 128.
1978 *Chamaeleo gastrotaenia* – BRYGOO, Faune de Madagascar, **47**: 172.
1986 *Calumma gastrotaenia* – KLAVER & BÖHME, Bonn. zool. Monogr., **22**: 58.

Distribution: Central and E. Madagascar.

Calumma gastrotaenia gastrotaenia (BOULENGER)

1900 *Chamaeleon grandidieri* MOCQUARD, Bull. Mus. Hist. nat. Paris, **6**: 345/Bull. Soc. philom., Paris, **2**: 96; plate 2, fig. 1. – Terra typica: Environs de Suberbieville, Madagascar. – Holotype (by monotypy): MHNP 99.416. – Syn.: fide WERNER 1902: 408.
1911 *Chamaeleon gastrotaenia* – WERNER, Das Tierreich, **23**: 26; fig. 7.
1966 *Chamaeleo gastrotaenia* – MERTENS, Das Tierreich, **83**: 15.
1970 *Chamaeleo gastrotaenia gastrotaenia* – BRYGOO, BLANC & DOMERGUE, Ann. Univ. Madagascar, Antananarivo, **7**: 275; fig.
1971 *Chamaeleo gastrotaenia gastrotaenia* – BRYGOO, Faune de Madagascar, **33**: 128; figs. 37–39, map 13.
1978 *Chamaeleo gastrotaenia gastrotaenia* – BRYGOO, Faune de Madagascar, **47**: 132.

Distribution: Central and E. Madagascar.

Calumma gastrotaenia andringitraensis (BRYGOO, BLANC & DOMERGUE)

1972 *Chamaeleo gastrotaenia andringitraensis* BRYGOO, BLANC & DOMERGUE, Bull. Mus. Hist. nat. Paris, **56** (Zool. **42**): 608; figs. 5–7. – Terra typica: Le long de la rivière Ampanasana à Ambalamarovandana, massif de l'Andringitra (1530 m alt.), Madagascar. – Holotype (by orig. design.): MHNP A330.
1978 *Chamaeleo gastrotaenia andringitraensis* – BRYGOO, Faune de Madagascar, **47**: 133; fig. 66.

Distribution: Andringitra massif, Madagascar.

Calumma gastrotaenia guillaumeti (BRYGOO, BLANC & DOMERGUE)

1974 *Chamaeleo gastrotaenia guillaumeti* BRYGOO, BLANC & DOMERGUE, Bull. Acad. malgache, Antananarivo (1973), **51**: 154; figs. 4–6. – Terra typica: Sommet du Marojezy (1300 m alt.), Madagascar. – Holotype (by orig. design.): MHNP 1973.442.
1978 *Chamaeleo gastrotaenia guillaumeti* – BRYGOO, Faune de Madagascar, **47**: 136; fig. 67.

Distribution: Marojezy massif and Tsaratanana (fide Ramanantsoa 1974), N. Madagascar.

Calumma gastrotaenia marojezensis (BRYGOO, BLANC & DOMERGUE)

1970 *Chamaeleo gastrotaenia marojezensis* BRYGOO, BLANC & DOMERGUE, Ann. Univ. malgache, Antananarivo, (Sci.), **7**: 278; 3 figs. – Terra typica: Massif du Marojezy (700 m alt.), Madagascar. – Holotype (by orig. design.): MHNP A160.
1971 *Chamaeleo gastrotaenia marojezensis* – BRYGOO, Faune de Madagascar, **33**: 136; fig. 40, map 13.
1978 *Chamaeleo gastrotaenia marojezensis* – BRYGOO, Faune de Madagascar, **47**: 138; fig. 68.

Distribution: Marojezy massif, Madagascar.

Calumma globifer (GÜNTHER)

1879 *Chamaeleon globifer* GÜNTHER, Proc. zool. Soc. London, **1879**: 149; plate 13. – Terra typica: Near Antananarivo, Madagascar. – Syntypes: BMNH 1946.8.21.53-54 (2 specimens).
1911 *Chamaeleon globifer* – WERNER, Das Tierreich, **27**: 30.
1966 *Chamaeleo globifer* – MERTENS, Das Tierreich, **83**: 16.
1971 *Chamaeleo globifer* – BRYGOO, Faune de Madagascar, **33**: 139; figs. 41–43, map 7.
1978 *Chamaeleo globifer* – BRYGOO, Faune de Madagascar, **47**: 139; figs. 69–70.
1986 *Calumma globifer* – KLAVER & BÖHME, Bonn. zool. Monogr., **22**: 58.

Distribution: E. Madagascar.

Calumma guibei (HILLENIUS)

1959 *Chamaeleo guibei* HILLENIUS, Beaufortia, Amsterdam, **8**: 30; figs. 22–23. – Terra typica: Mt. Tsaratanana (1800 m alt.), N.E. Madagascar. – Holotype (by orig. design.): MHNP 50.354.
1966 *Chamaeleo guibei* – MERTENS, Das Tierreich, **83**: 17.
1971 *Chamaeleo guibei* – BRYGOO, Faune de Madagascar, **33**: 145; fig. 44.
1978 *Chamaeleo guibei* – BRYGOO, Faune de Madagascar, **47**: 172.
1986 *Calumma guibei* – KLAVER & BÖHME, Bonn. zool. Monogr., **22**: 58.

Distribution: Known only from the type locality.

Calumma linota (MÜLLER)

1924 *Chamaeleon linotus* MÜLLER, Mitt. zool. Mus. Berlin, **11**: 95. – Terra typica: Madagascar. – Holotype (by orig. design.): ZSM 21/1923.
1966 *Chamaeleo linotus* – MERTENS, Das Tierreich, **83**: 19.
1971 *Chamaeleo linotus* – BRYGOO, Faune de Madagascar, **33**: 166; fig. 55.
1978 *Chamaeleo linotus* – BRYGOO, Faune de Madagascar, **47**: 172.
1986 *Calumma linota* – KLAVER & BÖHME, Bonn. zool. Monogr., **22**: 58.

Distribution: N.E. Madagascar.

Calumma malthe (GÜNTHER)

1879 *Chamaeleon malthe* GÜNTHER, Proc. zool. soc. London, **1879**: 148; plate 11. – Terra typica: Near Antananarivo, Madagascar. – Holotype (by monotypy): BMNH 1946.8.21.47.
1911 *Chamaeleon malthe* – WERNER, Das Tierreich, **27**: 32.
1966 *Chamaeleo malthe* – MERTENS, Das Tierreich, **83**: 19.
1971 *Chamaeleo malthe* – BRYGOO, Faune de Madagascar, **33**: 169; figs. 56–60, map 12.
1978 *Chamaeleo malthe* – BRYGOO, Faune de Madagascar, **47**: 142.
1986 *Calumma malthe* – KLAVER & BÖHME, Bonn. zool. Monogr., **22**: 58.

Distribution: N. and E. Madagascar.

Calumma nasuta (DUMÉRIL & BIBRON)

1836 *Camaeleon nasutus* DUMÉRIL & BIBRON, Erpétol. gén., **3**: 216. – Terra typica: Madagascar. – Syntypes: MHNP 6643 and 6643a–c (4 specimens).
1911 *Chamaeleon nasutus* – WERNER, Das Tierreich, **27**: 39.
1933 *Chamaeleo radamanus* MERTENS, Senckenbergiana, **15**: 269; fig. 3B–C. – Terra typica: Col Pierre Radama (1000 m alt.), Maroantsetra, N.E. Madagascar. – Holotype (by orig. design.): SMF 22132. – Syn.: fide ANGEL 1942: 169.
1966 *Chamaeleo nasutus* – MERTENS, Das Tierreich, **83**: 21.
1971 *Chamaeleo nasutus* – BRYGOO, Faune de Madagascar, **33**: 184; figs. 63–66, map 15.
1974 *Chamaeleo nasatus* PARCHER (ex errore), Z. Tierpsychol., **34**: 503.
1978 *Chamaeleo nasutus* – BRYGOO, Faune de Madagascar, **47**: 147; fig. 74.
1986 *Calumma nasuta* – KLAVER & BÖHME, Bonn. zool. Monogr., **22**: 58.

Distribution: Nosy Bé(?), Nosy Boraha and E. Madagascar.

Calumma oshaughnessyi (GÜNTHER)*

1881 *Chamaeleon o'shaughnessii* GÜNTHER (ex errore), Ann. Mag. nat. Hist., London, **7**: 358; plate 19. – Terra typica: Betsileo, Madagascar. – Holotype (by monotypy): BMNH 1946.8.21.88.
1978 *Chamaeleo oshaughnessyi* – BRYGOO, Faune de Madagascar, **47**: 172.
1986 *Calumma oshaughnessyi* – KLAVER & BÖHME, Bonn. zool. Monogr., **22**: 58.

Distribution: Madagascar.

Calumma oshaughnessyi oshaughnessyi (GÜNTHER)

1911 *Chamaeleon oshaughnessyi* – WERNER, Das Tierreich, **27**: 31.
1966 *Chamaeleo oshaughnessyi* – MERTENS, Das Tierreich, **83**: 21.
1971 *Chamaeleo oshaughnessyi* – BRYGOO, Faune de Madagascar, **33**: 192; figs. 67–70, map 7.
1974 *Chamaeleo oshaughnessyi oshaughnessyi* – RAMANANTSOA, Terre malgache, Antananarivo, **16**: 249.
1978 *Chamaeleo oshaughnessyi oshaughnessyi* – BRYGOO, Faune de Madagascar, **47**: 148; fig. 75.

Distribution: E. Madagascar.

Calumma oshaughnessyi ambreensis (RAMANANTSOA)

1974 *Chamaeleo oshaughnessyi ambreensis* RAMANANTSOA, Terre malgache, Anatananarivo **16**: 239; figs. 1–4. – Terra typica: Station forestière des Rousettes, Montagne d'Ambre (1000 m alt.), N. Madagascar. – Holotype (by orig. design.): MHNP 1974.6.
1978 *Chamaeleo oshaughnessyi ambreensis* – BRYGOO, Faune de Madagascar, **47**: 150.

Distribution: Known only from the type locality.

Calumma parsonii (CUVIER)

1824 *Chamael[eo] parsonii* CUVIER, Rech. Ossem. foss., 2nd ed., **5** (2): 269; plate XVI, figs. 30–31. – Terra typica: not given. – Type(s): MHNP, presumed lost (fide BRYGOO 1983: 4); based on CUVIER (1824): plate XVI, figs. 30–31.
1971 *Chamaeleo parsonii* – BRYGOO, Faune de Madagascar, **33**: 223.
1978 *Chamaeleo parsonii* – BRYGOO, Faune de Madagascar, **47**: 151.
1986 *Calumma parsonii* – KLAVER & BÖHME, Bonn. zool. Monogr., **22**: 58.

Distribution: Nosy Boraha and Madagascar.

Calumma parsonii parsonii (CUVIER)

1832 *Chamaeleo madecasseus* LESSON, Illustr. Zool.: plate 35. – Terra typica: Madagascar – Syntype: MHNP 8177 (the only remaining specimen of the syntype series according to BRYGOO (1983: 13). – Syn.: fide DUMÉRIL & BIBRON 1836: 231.
1843 *Chamaeleo madecassus* FITZINGER (ex errore pro *Chamaeleo madacasseus* LESSON 1832), Syst. Rept.: 42.
1887 *Chamaeleon parsonsii* BOULENGER (nomen emendatum pro *Chamaeleo parsonii* CUVIER 1824), Cat. Liz. Brit. Mus., **3**: 466.
1893 *Chamaeleon personii* MOCQUARD (ex errore), C.R. Soc. philom., **19**: 6.
1911 *Chamaeleon parsonii* – WERNER, Das Tierreich, **27**: 31.
1913 *Chamaeleon parsonsi parsonsi* METHUEN & HEWITT (ex errore), Ann. Transvaal. Mus. Pretoria, **3**: 191.
1922 *Chamaeleon partonsi* KAUDERN (ex errore), Zool. Jb., Syst., **45**: 431.
1966 *Chamaeleo parsonii* – MERTENS (partim), Das Tierreich, **83**: 22.
1971 *Chamaeleo parsonii parsonii* – BRYGOO, Faune de Madagascar, **33**: 223; figs. 79–82, map 18.
1978 *Chamaeleo parsonii parsonii* – BRYGOO, Faune de Madagascar, **47**: 172.

Distribution: Nosy Boraha and Madagascar.

Calumma parsonii cristifer (METHUEN & HEWITT)

1913 *Chamaeleon parsonsi* var. *cristifer* METHUEN & HEWITT, Ann. Transvaal Mus., Pretoria, **3**: 191; plate 11, figs. 1–3. – Terra typica: Analamazotra, Madagascar. – Syntypes: TM 4222–27 (6 specimens).
1966 *Chamaeleo parsonii* – MERTENS (partim), Das Tierreich, **83**: 22.
1971 *Chamaeleo parsonii* var. *cristifer* – BRYGOO, Faune de Madagascar, **33**: 234; figs. 83–85, map 18.
1973 *Chamaeleo parsonii cristifer* – MERTENS, Salamandra, **9**: 56.
1978 *Chamaeleo parsonii* var. *cristifer* – BRYGOO, Faune de Madagascar, **47**: 151; fig. 76.
1988 *Chamaeleo parsoni crucifer* TELFORD (ex errore), Internat. J. Parasitol., **18**: 197.

Distribution: Périnet, Madagascar.

Calumma peyrierasi (BRYGOO, BLANC & DOMERGUE)

1974 *Chamaeleo peyrierasi* BRYGOO, BLANC & DOMERGUE, Bull. Acad. malgache, Antananarivo (1973), **51**: 151; figs. 1–3. – Terra typica: Flancs du Marojezy (1900–2000 m alt.), N. Madagascar. – Holotype (by orig. design.): MHNP 1973.440.
1978 *Chamaeleo peyrierasi* – BRYGOO, Faune de Madagascar, **47**: 153; fig. 77.
1986 *Calumma peyrierasi* – KLAVER & BÖHME, Bonn. zool. Monogr., **22**: 58.

Distribution: Known only from the type locality.

Calumma tigris (KUHL)

1820 *Chamaeleo tigris* KUHL, Beitr. Zool. vergl. Anat., **1**: 104. – Terra typica: not given; according to GUIBÉ (1954: 105): Iles Seychelles. – Holotype (by monotypy): MHNP 6640.
1820 *Chamaeleo scychellensis* KUHL, Beitr. Zool. vergl. Anat., **1**: 105. – Terra typica: In insulis Scychelles dictis. – Syntype: MHNP 220 (the only remaining specimen of the syntype series according to BRYGOO 1983: 17). – Syn.: fide DUMÉRIL & BIBRON 1836: 212.
1829 *Chamaeleo seichellensis* CUVIER (nomen emendatum pro *Chamaeleo scychellensis* KUHL 1820), Règne animal, **2**: 60; note 1.
1836 *Chamaeleo seychellensis* DUMÉRIL & BIBRON (nomen emendatum pro *Chamaeleo scychellensis* KUHL 1820), Erpétol. gén., **3**: 212.

1843 *Chamaeleon scychelliensis* FITZINGER (nomen emendatum pro *Chamaeleo scychellensis* KUHL 1820), Syst. Rept.: 42.
1845 *Chameleo fimbriatus* GRAY (nomen nudum), Cat. Lizards. Brit. Mus.: 268.
1911 *Chamaeleon tigris* – WERNER, Das Tierreich, **27**: 24; fig. 6.
1911 *Chamaeleon sychellensis* WERNER (ex errore), Das Tierreich, **27**: 24.
1966 *Chamaeleo tigris* – MERTENS, Das Tierreich, **83**: 28.
1978 *Chamaeleo tigris* – BRYGOO, Faune de Madagascar, **47**: 167; figs. 89–91.
1986 *Calumma tigris* – KLAVER & BÖHME, Bonn. zool. Monogr., **22**: 58.

Distribution: Islands Mahé, Silhouette and Praslin, Seychelles; introduced on Zanzibar (fide BOULENGER 1887: 461 & WERNER 1911: 24).

Calumma tsaratananensis (BRYGOO & DOMERGUE)*

1968 *Chamaeleo tsaratananensis* BRYGOO & DOMERGUE, Bull. Mus. Hist. nat. Paris (1967), **39**: 832; figs. 1–2. – Terra typica: Massif of Tsaratanana (2500 m alt.), N. Madagascar. – Holotype (by orig. design.): MHNP A59.
1971 *Chamaeleo tsaratananensis* – BRYGOO, Faune de Madagascar, **33**: 247; fig. 89, map 12.
1978 *Chamaeleo tsaratananensis* – BRYGOO, Faune de Madagascar, **47**: 172.
1986 *Calumma tsaratananensis* – KLAVER & BÖHME, Bonn. zool. Monogr., **22**: 58.

Distribution: Known only from the type locality.

Genus **Furcifer** FITZINGER*

1839 *Diceros* SWAINSON (non *Diceros* GRAY 1821 = Mammalia, non *Diceros* BURMEISTER 1842 = Coleoptera), Nat. Hist. Fishes Amph. Rept., **2**: 369. – Species typica (by monotypy): *Diceros bifurcatus* SWAINSON (= *Chamaeleo bifidus* BRONGNIART 1800).
1843 *Furcifer* FITZINGER (non *Furcifer* WAGNER 1844 = Mammalia), Syst. Rept.: 15, 42. – Species typica (by original designation): *Chamaeleon bifidus* BRONGNIART 1800.
1865 *Apola* GRAY, Proc. zool. Soc. London, **1864**: 467, 473; fig. – Species typica (by monotypy): *Chamaeleo lateralis* GRAY 1831.
1865 *Sauroceras* GRAY, Proc. zool. Soc. London, **1864**: 468, 478. – Species typica (by monotypy): *Chameleo rhinoceratus* GRAY 1845.
1865 *Dicranosaura* GRAY, Proc. zool. Soc. London, **1864**: 468, 478. – Species typica (by subsequent designation: LOVERIDGE 1957a: 197): *Chameleo bifurcus* GRAY (= *Chamaeleo bifidus* BRONGNIART 1800).
1865 *Cyneosaura* GRAY, Proc. zool. Soc. London, **1864**: 468, 479. – Species typica (by monotypy): *Chameleo pardalis* GRAY (= CUVIER 1829).

Distribution: Madagascar, adjacent islets and Comores.

Furcifer angeli (BRYGOO & DOMERGUE)*

1968 *Chamaeleo angeli* BRYGOO & DOMERGUE, Mém. Mus. Hist. nat. Paris, **52**: 104; figs. 19–21. – Terra typica: km 450 de la route nationale 4, N. of Tsaramandroso, forêt de l'Ankarafantsika, Madagascar. – Holotype (by orig. design.): MHNP 1967.174.
1971 *Chamaeleo angeli* – BRYGOO, Faune de Madagascar, **33**: 39; figs. 1–2, map 6.
1978 *Chamaeleo angeli* – BRYGOO, Faune de Madagascar, **47**: 112; fig. 52.
1986 *Furcifer angeli* – KLAVER & BÖHME, Bonn. zool. Monogr., **22**: 58.

Distribution: N.W. Madagascar.

Furcifer antimena (GRANDIDIER)

1872 *Chamaeleo antimena* GRANDIDIER, Ann. Sci. nat., Zool., Paris, **15**, art. 20: 7 / Bibl.Ecole Hautes Etudes, **6**: 7. – Terra typica: Côte occidentale de Madagascar. – Holotype (by monotypy): MHNP 5467.
1911 *Chamaeleon rhinoceratus* – WERNER (partim), Das Tierreich, **27**: 21.
1913 *Chamaeleon rhinoceratus lineatus* METHUEN & HEWITT, Ann. Transvaal. Mus., Pretoria, **3**: 190; plate 10. – Terra typica: Tsivanóa, Madagascar. – Holotype (by monotypy): TM 4200. – Syn.: fide BRYGOO & DOMERGUE 1968: 85).
1942 *Chamaeleo antinema* ANGEL (ex errore), Mém. Acad. malgache, Antananarivo, **36**: 165.
1966 *Chamaeleo rhinoceratus rhinoceratus* – MERTENS (partim), Das Tierreich, **83**: 25.
1968 *Chamaeleo antimena* – BRYGOO & DOMERGUE, Mém. Mus. Hist. nat. Paris, (A, zool.), **52**: 84; figs. 6–11, map.
1971 *Chamaeleo antimena* – BRYGOO, Faune de Madagascar, **33**: 45; figs. 3–6, map 6.
1978 *Chamaeleo antimena* – BRYGOO, Faune de Madagascar, **47**: 114.
1986 *Furcifer antimena* – KLAVER & BÖHME, Bonn. zool. Monogr., **22**: 58.

Distribution: S.W. Madagascar.

Furcifer balteatus (DUMÉRIL & BIBRON)*

1851 *Chamaeleo balteatus* DUMÉRIL & BIBRON in: DUMÉRIL & DUMÉRIL, Cat. méthod. Coll. Rept.: 32. – Terra typica: Madagascar. – Holotype (by monotypy): MHNP 6521.
1911 *Chamaeleon bifidus* – WERNER (partim), Das Tierreich, **27**: 27.
1942 *Chamaeleon balteus* ANGEL (ex errore), Mém. Acad. malgache, Antananarivo, **36**: 174.
1966 *Chamaeleo bifidus* – MERTENS (partim), Das Tierreich, **83**: 7.
1969 *Chamaeleo balteatus* – BRYGOO & DOMERGUE, Bull. Mus. Hist. nat. Paris, **41**: 105; figs. 1–8.
1971 *Chamaeleo balteatus* – BRYGOO, Faune de Madagascar, **33**: 54; figs. 7–9, map 7.
1978 *Chamaeleo balteatus* – BRYGOO, Faune de Madagascar, 47: 114.
1986 *Furcifer balteatus* – KLAVER & BÖHME, Bonn. zool. Monogr., **22**: 58.

Distribution: E. Madagascar.

Furcifer belalandaensis (BRYGOO & DOMERGUE)*

1970 *Chamaeleo belalandaensis* BRYGOO & DOMERGUE, Bull. Mus. Hist. nat. Paris, **42**: 309; figs. 1–3. – Terra typica: Village de Belalanda, 4 km N. of Tuléar, S.W. Madagascar. – Holotype (by orig. design.): MHNP 1969.114.
1971 *Chamaeleo belalandaensis* – BRYGOO, Faune de Madagascar, **33**: 63; fig. 10, map 11.
1978 *Chamaeleo belalandaensis* – BRYGOO, Faune de Madagascar, **47**: 172.
1986 *Furcifer belalandaensis* – KLAVER & BÖHME, Bonn. zool. Monogr., **22**: 58.

Distribution: Known only from the type locality.

Furcifer bifidus (BRONGNIART)

1800 *Chamaeleo bifidus* BRONGNIART, Bull. Soc. philom., Paris, **2**: 90; plate 6, fig. 2. – Terra typica: not given. – Holotype (by monotypy): MHNP 6660.
1820 *Chamaeleo bifurcus* KUHL (ex errore), Beitr. Zool. vergl. Anat., **1**: 103.
1839 *Diceros bifurcatus* SWAINSON (nomen emendatum pro *Chamaeleo bifurcus* KUHL 1820), Nat. Hist. Fishes Amph. Rept., **2**: 369.
1843 *Chamaeleo (Furcifer) brongniartii* FITZINGER (nomen substitutum pro *Chamaeleo bifidus* BRONGNIART 1800), Syst. Rept.: 42.
1865 *Dicranosaura bifurca* var. *crassicornis* GRAY, Proc. zool. Soc. London, **1864**: 479. – Terra typica: Madagascar. – Type(s): BMNH; not registered as type(s). – Syn.: fide BOULENGER, 1887: 463.

1911 *Chamaeleon bifidus* – WERNER (partim), Das Tierreich, **27**: 27.
1966 *Chamaeleo bifidus* – MERTENS (partim), Das Tierreich, **83**: 7.
1971 *Chamaeleo bifidus* – BRYGOO, Faune de Madagascar, **33**: 65; figs. 11–14, map. 8.
1971 *Dicranosaura bifurca* var. *crassirostris* BRYGOO (ex errore pro *Dicranosaura bifurca* var. *crassicornis* GRAY 1865), Faune de Madagascar, **33**: 65.
1978 *Chamaeleo bifidus* – BRYGOO, Faune de Madagascar **47**: 116; figs. 53–54.
1986 *Furcifer bifidus* – KLAVER & BÖHME, Bonn. zool. Monogr., **22**: 58.

Distribution: E. Madagascar.

Furcifer campani (GRANDIDIER)

1872 *Chamaeleon campani* GRANDIDIER, Ann. Sci. nat., Zool., Paris, **15**, art. 20: 8 / Bibl. Ecole Hautes Etudes, **6**: 8. – Terra typica: Massif de l'Ankaratra, Madagascar. – Holotype (by monotypy): MHNP 5465.
1881 *Chamaeleo octotaeniatus* BOETTGER, Abh. senckenb. naturf. Ges., **12**: 481. – Terra typica: Tsiafakafo (zwischen 19°–20° südl. Breite und 47°–48° östl. Länge), Madagascar. – Holotype (by monotypy): UMB; presumed lost (H. Hohmann, pers. comm.). – Syn.: fide BOULENGER, 1887: 465.
1911 *Chamaeleon campani* – WERNER, Das Tierreich, **27**: 17.
1966 *Chamaeleo campani* – MERTENS, Das Tierreich, **83**: 9.
1971 *Chamaeleo campani* – BRYGOO, Faune de Madagascar, **33**: 97; figs. 24–25, map 11.
1978 *Chamaeleo campani* – BRYGOO, Faune de Madagascar, **47**: 126.
1986 *Furcifer campani* – KLAVER & BÖHME, Bonn. zool. Monogr., **22**: 58.

Distribution: Central Madagascar.

Furcifer cephalolepis (GÜNTHER)

1880 *Chamaeleon cephalolepis* GÜNTHER, Ann. Mag. nat. Hist., London, **6**: 237; fig. – Terra typica: Comoro islands. – Syntypes: BMNH 1946.8.22.28–29 (2 specimens).
1911 *Chamaeleon cephalolepis* – WERNER, Das Tierreich, **27**: 20.
1966 *Chamaeleo cephalolepis* – MERTENS, Das Tierreich, **83**: 9.
1971 *Chamaeleo cephalolepis* – BRYGOO, Faune de Madagascar, **33**: 282; figs. 106–107.
1978 *Chamaeleo cephalolepis* – BRYGOO, Faune de Madagascar, **47**: 164; figs. 85–86.
1986 *Furcifer cephalolepis* – KLAVER & BÖHME, Bonn. zool. Monogr., **22**: 58.

Distribution: Grande Comore, Comores.

Furcifer labordi (GRANDIDIER)

1872 *Chamaeleo labordi* GRANDIDIER, Ann. Sci. nat., Zool., Paris, **15** (art. 20): 7 / Bibl. Ecole Hautes Etudes **6**: 7. – Terra typica: Côte occidentale de Madagascar. – Syntypes: MHNP 5469 and 5470 (3 specimens).
1911 *Chamaeleon labordi* – WERNER, Das Tierreich, **27**: 21.
1942 *Chamaeleo barbouri* HECHENBLEIKNER, Proc. New Engl. zool. Club, Boston, **19**: 103. – Terra typica: Madagascar. – Holotype (by orig. design.): MCZ 7758. – Syn.: fide HILLENIUS 1959: 13.
1966 *Chamaeleo rhinoceratus labordi* – MERTENS, Das Tierreich, **83**: 25.
1968 *Chamaeleo labordi* – BRYGOO & DOMERGUE, Mém. Mus. Hist. nat., Paris, (A, Zool.), **52**: 77; figs. 1–5, 11, map.
1971 *Chamaeleo labordi* – BRYGOO, Faune de Madagascar, **33**: 147; figs. 45–49, map 6.
1978 *Chamaeleo labordi* – BRYGOO, Faune de Madagascar, **47**: 140.
1986 *Furcifer labordi* – KLAVER & BÖHME, Bonn. zool. Monogr., **22**: 58.

Distribution: W. Madagascar.

Furcifer lateralis (GRAY)*

1831 *Chamaeleo lateralis* GRAY in GRIFFITH, Anim. Kindgd. **9** (Synops.): 53. – Terra typica: not given, according to GRAY (1845: 264): Madagascar. – Syntypes: BMNH 1946.8.22.12–13 (2 specimens).
1911 *Chamaeleon lateralis* – WERNER, Das Tierreich, **27**: 16.
1921 *Chamaeleo lambertoni* ANGEL, Bull. Must. Hist. nat., Paris, **27**: 411; fig. 7. – Terra typica: Tananarive, Madagascar. – Holotype (by monotypy): MHNP 1921.269. – Syn.: fide HILLENIUS 1959: 17.
1966 *Chamaeleo lateralis* – MERTENS, Das Tierreich, **83**: 19.
1971 *Chamaeleo lateralis* – BRYGOO, Faune de Madagascar, **33**: 156; figs. 50–54, map 14.
1971 *Chamaeleo lateralis major* BRYGOO, Faune de Madagascar, **33**: 164; fig. 51, map 14. – Terra typica: Tanandava, Madagascar. – Holotype (by implication): MHNP 457/C. – Syn.: hoc loco.
1978 *Chamaeleo lateralis* – BRYGOO, Faune de Madagascar, **47**: 141.
1986 *Furcifer lateralis* – KLAVER & BÖHME, Bonn. zool. Monogr., **22**: 58.
1994 *C[hamaeleo] parietalis* NEČAS (nomen nudum), Herpetozoa, Vienna, **7**: 100.

Distribution: Madagascar.

Furcifer minor (GÜNTHER)

1879 *Chamaeleon minor* GÜNTHER, Ann. Mag. nat. Hist., London, **4**: 246; plate 13. – Terra typica: Fianarantsoa, Betsileo, S.E. Madagascar. – Syntypes: BMNH 1946.8.21.42–45 (4 specimens).
1911 *Chamaeleon minor* – WERNER, Das Tierreich, **27**: 27.
1966 *Chamaeleo minor* – MERTENS, Das Tierreich, **83**: 20.
1971 *Chamaeleo minor* – BRYGOO, Faune de Madagascar, **33**: 179; fig. 61, map 8.
1978 *Chamaeleo minor* – BRYGOO, Faune de Madagascar, **47**: 144; figs. 72–73.
1986 *Furcifer minor* – KLAVER & BÖHME, Bonn. zool. Monogr., **22**: 58.

Distribution: S. Madagascar.

Furcifer monoceras (BOETTGER)

1913 *Chamaeleon monoceras* BOETTGER in VOELTZKOW, Reise Ostafr., Rept.Amph.: 307; plate 26, fig. 3. – Terra typica: Betsako bei Mojunga, N.W. Madagascar. – Holotype (by monotypy: SMF 16420.
1966 *Chamaeleo monoceras* – MERTENS, Das Tierreich, **83**: 20.
1971 *Chamaeleo monoceras* – BRYGOO, Faune de Madagascar, **33**: 182; fig. 62.
1978 *Chamaeleo monoceras* – BRYGOO, Faune de Madagascar, **47**: 146.
1986 *Furcifer monoceras* – KLAVER & BÖHME, Bonn. zool. Monogr., **22**: 58.

Distribution: N.W. Madagascar.

Furcifer oustaleti (MOCQUARD)*

1894 *Chamaeleon oustaleti* MOCQUARD, C.R. Soc. philom., Paris, **9**: 3. – Terra typica: Betsileo and Diégo Suarez, Madagascar. – Syntypes: MHNP 1893. 155–158 and 1891.22 (5 specimens).
1911 *Chamaeleon oustaleti* – WERNER, Das Tierreich, **27**: 31.
1966 *Chamaeleo oustaleti* – MERTENS, Das Tierreich, **83**: 21.
1971 *Chamaeleo oustaleti* – BRYGOO, Faune de Madagascar, **33**: 201; figs. 71–73, map 16.
1978 *Chamaeleo oustaleti* – BRYGOO, Faune de Madagascar, **47**: 150.
1986 *Furcifer oustaleti* – KLAVER & BÖHME, Bonn. zool. Monogr., **22**: 58.

Distribution: Madagascar, introduced in East Africa: Ngong forest, Nairobi, Kenya (fide LOVERIDGE 1957a: 206).

Furcifer pardalis (CUVIER)*

1829 *Chamaeleo pardalis* CUVIER, Règne. animal, 2nd ed., **2**: 60. – Terra typica: Ile de France (= Mauritius). – Holotype (by monotypy): MHNP 6520.
1832 *Chamaeleo ater* LESSON, Illustr. Zool.: plate 34. – Terra typica: Madagascar. – Type(s): MHNP, presumed lost (fide BRYGOO 1983: 4). – Syn.: fide DUMÉRIL & BIBRON 1836: 228.
1836 *Chamaeleo niger* DUMÉRIL & BIBRON (ex errore pro *Chamaeleo ater* LESSON 1832), Erpétol. gén. **3**: 228.
1888 *Chamaeleo guentheri* BOULENGER, Ann. Mag. nat. Hist., London, **1**: 22; plate 2, figs. 1–2. – Terra typica: Nossi Bé, near Madagascar. – Syntypes: BMNH 1946.8.22.21–25 (5 specimens). – Syn.: fide BRYGOO 1969: 117.
1891 *Chamaeleon longicauda* GÜNTHER, Ann. Mag. nat. Hist., London, **8**: 287; plate 14. – Terra typica: Anorontsangana, N.W. Madagascar. – Holotype (by monotypy): BMNH 1946.8.21.93. – Syn.: fide MOCQUARD 1909: 6.
1899 *Chamaeleon axillaris* WERNER, Zool. Anz., **22**: 183. – Terra typica: Madagascar. – Syntypes: not located (2 specimens). – Syn.: fide WERNER 1902: 377.
1911 *Chamaeleon pardalis* – WERNER, Das Tierreich, **27**: 19.
1911 *Chamaeleon guentheri* – WERNER, Das Tierreich, **27**: 20.
1911 *Chamaeleon longicauda* – WERNER, Das Tierreich, **27**: 20.
1923 *Chamaeleon krempfi* CHABANAUD, Bull. Mus. Hist. nat., Paris, **29**: 209. – Terra typica: L'une des petits îles du golfe de Siam, peut-être Poulo Condore. – Terra typica designata (fide MERTENS 1966: 22): Madagascar. – Syntypes: MHNP 1923.26–27 (2 specimens). – Syn.: fide CHABANAUD 1923: 558.
1966 *Chamaeleo pardalis* – MERTENS, Das Tierreich, **83**: 20.
1966 *Chamaeleo guentheri* – MERTENS, Das Tierreich, **83**: 17.
1971 *Chamaeleo pardalis* – BRYGOO, Faune de Madagascar, **33**: 210; figs, 74–78, map 17.
1978 *Chamaeleo pardalis* – BRYGOO, Faune de Madagascar, **47**: 150.
1986 *Furcifer pardalis* – KLAVER & BÖHME, Bonn. zool. Monogr., **22**: 59.

Distribution: Nosy Bé, Nosy Boraha, Nosy Faly, Madagascar and introduced on Réunion (fide BOURGAT 1970: 260).

Furcifer petteri (BRYGOO & DOMERGUE)

1966 *Chamaeleo willsi petteri* BRYGOO & DOMERGUE, Bull. Mus. Hist. nat. Paris, **38**: 361. – Terra typica: Bordure Est du massif de l'Ankarana, Madagascar. – Holotype (by orig. design.): MHNP A47.
1971 *Chamaeleo willsii petteri* – BRYGOO, Faune de Madagascar, **33**: 270; figs. 99–101, map 20.
1978 *Chamaeleo willsii petteri* – BRYGOO, Faune de Madagascar, **47**: 162.
1978 *Chamaeleo petteri* – RAMANANTSOA, Bull. Mus. Hist. nat. Paris, (Zool.) **351**: 101; figs. 6–7.
1986 *Furcifer petteri* – KLAVER & BÖHME, Bonn. zool. Monogr., **22**: 59.

Distribution: N. Madagascar.

Furcifer polleni (PETERS)

1874 *Chamaeleo polleni* PETERS, M.Ber. Akad. Wiss. Berlin, **1873**: 792. – Terra typica: Mayotte, Comoren. – Holotype (by monotypy): ZMB 7892.
1911 *Chamaeleon polleni* – WERNER, Das Tierreich, **27**: 21.
1966 *Chamaeleo polleni* – MERTENS, Das Tierreich, **83**: 23.
1971 *Chamaeleo polleni* – BRYGOO, Faune de Madagascar, **33**: 286; figs. 108–109.
1978 *Chamaeleo polleni* – BRYGOO, Faune de Madagascar, **47**: 165; figs. 87–88.
1986 *Furcifer polleni* – KLAVER & BÖHME, Bonn. zool. Monogr., **22**: 59.

Distribution: Mayotte, Comores.

Furcifer rhinoceratus (GRAY)

1843 *Chamelo rhinoceros* GRAY (nomen dubium), Ann. Mag. nat. Hist., London, **9**: 46. – Terra typica: Madagascar?
1845 *Chameleo rhinoceratus* GRAY, Cat. Lizards. Brit. Mus.: 267. – Terra typica: Madagascar. – Holotype (by monotypy): BMNH 1946.8.20.98.
1893 *Chamaeleon voeltzkowi* BOETTGER, Kat. Rept. Mus. senckenb. naturf. Ges., **1**: 120, footnote. – Terra typica: Antema an der Bembatukabai, W. Madagascar. – Holotype (by orig. design.): SMF 16375. – Syn.: fide HILLENIUS 1959: 15.
1911 *Chamaeleon rhinoceratus* – WERNER (partim), Das Tierreich, **27**: 21.
1911 *Chamaeleon voeltzkowi* – WERNER, Das Tierreich, **27**: 21.
1966 *Chamaeleo rhinoceratus* – MERTENS, Das Tierreich, **83**: 25.
1966 *Chamaeleo rhinoceratus rhinoceratus* – MERTENS (partim), Das Tierreich, **83**: 25.
1966 *Chamaeleo rhinoceratus voeltzkowi* – MERTENS, Das Tierreich, **83**: 26.
1971 *Chamaeleo rhinoceratus* – BRYGOO, Faune de Madagascar, **33**: 239; figs. 86–88, map 6.
1978 *Chamaeleo rhinoceratus* – BRYGOO, Faune de Madagascar, **47**: 155; figs. 78–80.
1986 *Furcifer rhinoceratus* – KLAVER & BÖHME, Bonn. zool. Monogr., **22**: 59.

Distribution: N.W. Madagascar.

Furcifer tuzetae (BRYGOO, BOURGAT & DOMERGUE)

1972 *Chamaeleo tuzetae* BRYGOO, BOURGAT & DOMERGUE, Bull. Mus. Hist. nat. Paris, **27** (Zool., **21**): 139; figs. 1–3. – Terra typica: Andrenalamivola, canton de Befandriana sud, S.W. Madagascar. – Holotype (by monotypy): MHNP A258.
1978 *Chamaeleo tuzetae* – BRYGOO, Faune de Madagascar, **47**: 158; fig. 81.
1986 *Furcifer tuzetae* – KLAVER & BÖHME, Bonn. zool. Monogr., **22**: 59.

Distribution: Known only from the type locality.

Furcifer verrucosus (CUVIER)*

1829 *Chamaeleo verrucosus* CUVIER, Règne animal, 2nd ed., **2**: 60. – Terra typica: Ile Bourbon (= Réunion). – Holotype (by monotypy): MHNP 6635.
1831 *Chamaeleon monilifer* GRAY (nomen nudum) in GRIFFITH, Animal Kingdom, **9** (Synops.): 53. – Syn.: fide GRAY 1831: 53.
1966 *Chamaeleo verrucosus* – MERTENS, Das Tierreich, **83**: 28.
1971 *Chamaeleo verrucosus* – BRYGOO, Faune de Madagascar, **33**: 250, map 19.
1978 *Chamaeleo verrucosus* – BRYGOO, Faune de Madagascar, **47**: 161.
1986 *Furcifer verrucosus* – KLAVER & BÖHME, Bonn. zool. Monogr., **22**: 59.

Distribution: Madagascar.

Furcifer verrucosus verrucosus (CUVIER)

1911 *Chamaeleon verrucosus* – WERNER, DAS TIERREICH, **27**: 28.
1966 *Chamaeleo verrucosus verrucosus* – MERTENS, Das Tierreich, **83**: 28.
1971 *Chamaeleo verrucosus verrucosus* – BRYGOO, Faune de Madagascar, **33**: 250; figs. 90–93.
1978 *Chamaeleo verrucosus verrucosus* – BRYGOO, Faune de Madagascar, **47**: 172.

Distribution: Madagascar, except the south.

Furcifer verrucosus semicristatus (BOETTGER)

1894 *Chamaeleon semicristatus* BOETTGER, Zool. Anz., **17**: 138. – Terra typica: Südspitze von Madagascar. – Holotype (by monotypy): SMF 16371.
1911 *Chamaeleon semicristatus* – WERNER, Das Tierreich, **27**: 19.

1966 *Chamaeleo verrucosus semicristatus* – MERTENS, Das Tierreich, **83**: 28.
1971 *Chamaeleo verrucosus semicristatus* – BRYGOO, Faune de Madagascar, **33**: 258; fig. 94.
1978 *Chamaeleo verrucosus semicristatus* – BRYGOO, Faune de Madagascar, **47**: 172.

Distribution: S. Madagascar.

Furcifer willsii (GÜNTHER)

1890 *Chamaeleon willsii* GÜNTHER, Ann. Mag. nat. Hist., London, **5**: 71; plate 6. – Terra typica: Forest region W. of Imerina, Madagascar. – Syntypes: BMNH 1946.8.21.59–62 (4 specimens).
1911 *Chamaeleon willsi* – WERNER (ex errore), Das Tierreich, **27**: 29.
1966 *Chamaeleo willsii* – MERTENS, Das Tierreich, **83**: 29.
1971 *Chamaeleo willsii willsii* – BRYGOO, Faune de Madagascar, **33**: 263; figs. 95–98, map 20.
1978 *Chamaeleo willsii willsii* – BRYGOO, Faune de Madagascar, **47**: 162.
1978 *Chamaeleo willsii* – RAMANANTSOA, Bull. Mus. Hist. nat. Paris, (Zool.) **351**: 101; figs. 4 & 7.
1986 *Furcifer willsii* – KLAVER & BÖHME, Bonn. zool. Monogr., **22**: 59.

Distribution: Central Madagascar.

Genus **Bradypodion** FITZINGER*

1843 *Bradypodion* FITZINGER, Syst. Rept.: 15, 43. – Species typica (by original designation): *Chamaeleon pumilus* LATREILLE (= GMELIN 1789).
1846 *Bradypodium* AGASSIZ (nomen emendatum pro *Bradypodion* FITZINGER 1843), Nomencl. Zool., Index univers.: 52.
1865 *Microsaura* GRAY, Proc. zool. soc. London, **1864**: 467, 473; fig. – Species typica (by monotypy): *Microsaura melanocephala* GRAY 1865.
1865 *Lophosaura* GRAY (non *Lophosaura* GRAY 1852 = Reptilia, Iguanidae), Proc. zool. Soc. London, **1864**: 468, 474. – Species typica (by subsequent designation: LOVERIDGE 1957a: 197): *Lacerta pumila* GMELIN 1789.
1942 *Brachypodion* ANGEL (ex errore), Mém. Acad. malgache., Antananarivo, **36**: 175.
1962 *Microsaurus* VON FRISCH (ex errore; non *Microsaurus* FRAAS 1866 = Reptilia, Phytosauridae; non *Microsaurus* HATCHER 1900 = Reptilia, Sauropoda incertae sedis), Z. Tierpsychol., **19**: 276.

Distribution: Central, E. and S. Africa.

Bradypodion adolfifriderici (STERNFELD)

1912 *Chamaeleon adolfi-friderici* STERNFELD, Wiss. Ergebn. dtsch. Zentr. Afr.-Exped. 1907–1908, **4** (Zool. 2): 258. – Terra typica: Irumu-Mavambi forest, Ituri, Zaire. – Holotype (by monotypy): ZMB 22709.
1963 *Chamaeleo adolfi-friederici* HILLENIUS (ex errore), Beaufortia, Amsterdam, **9**: 208.
1966 *Chamaeleo adolfifriderici* – MERTENS, Das Tierreich, **83**: 6.
1986 *Bradypodion adolfifriderici* – KLAVER & BÖHME, Bonn. zool. Monogr., **22**: 59.

Distribution: Rwanda, Uganda and E. Zaire.

Bradypodion carpenteri (PARKER)

1929 *Chamaeleo carpenteri* PARKER, Ann. Mag. nat. Hist., London, **3**: 280; fig. – Terra typica: Ruwenzori Mts. (6500 ft. alt.), Uganda. – Holotype (by orig. design.): BMNH 1928.11.23.6.
1966 *Chamaeleo carpenteri* – MERTENS, Das Tierreich, **83**: 9.
1986 *Bradypodion carpenteri* – KLAVER & BÖHME, Bonn. zool. Monogr., **22**: 59.

Distribution: Uganda and E. Zaire.

Bradypodion dracomontanum Raw

1976 *Bradypodion dracomontanum* Raw, Durban Mus. Novitates, **11**: 157; fig. 4D, plate 1, map 1. – Terra typica: Cathedral Peak, Natal Drakensberg, South Africa. – Holotype (by orig. design.): NM 1168.

Distribution: Drakensberg Range, South Africa.

Bradypodion fischeri (Reichenow)*

1887 *Chamaeleon fischeri* Reichenow, Zool. Anz., **10**: 371. – Terra typica: Nguru Berge (Usagara), Tanganyika (= Tanzania). – Holotype (by monotypy): ZMB 10744.
1966 *Chamaeleo fischeri* – Mertens, Das Tierreich, **83**: 13.
1986 *Bradypodion fischeri* – Klaver & Böhme, Bonn. zool. Monogr., **22**: 59.

Distribution: Kenya and Tanzania.

Bradypodion fischeri fischeri (Reichenow)

1896 *Chamaeleon matschiei* Werner, Verh. zool.-bot. Ges. Wien (1895) **45**: 192. – Terra typica: Usambara-Berge, Tanganyika (= Tanzania). – Type(s): not located. – Syn.: fide Werner, 1902: 414.
1902 *Chamaeleon tornieri* Werner, Zool. Jb., Syst., **15**: 417; plate 24. – Terra typica: Mozambique. – Holotype (by monotypy): ZMB 17392. – Syn.: fide Barbour & Loveridge 1928: 172.
1911 *Chamaeleon fischeri* – Werner, Das Tierreich, **27**: 28.
1911 *Chamaeleon fischeri fischeri* – Werner, Das Tierreich, **27**: 28.
1911 *Chamaeleon fischeri matschiei* – Werner, Das Tierreich, **27**: 28.
1913 *Chamaeleon fischeri vosseleri* Nieden, S.Ber. Ges. naturf. Fr. Berlin, **1913**: 247; figs. 11–18; plate 15, fig. 5; plate 16, fig. 9. – Terra typica: Tanga, Buloa bei Tanga, Magrotto, Amani, Nguelo, Ukami and Usamaro, Tanganyika (= Tanzania). – Terra typica restricta (fide Loveridge 1957a: 204; footnote 46): Amani, Bulwa, Nguelo in Usambara Mts., Magrotto Mt., Tanga, Tanzania. – Syntypes: ZMB 14947 (1 specimen) and NMW 23381 (1 specimen). – Syn.: fide Barbour & Loveridge 1928: 172.
1966 *Chamaeleo fischeri fischeri* – Mertens, Das Tierreich, **83**: 14.

Distribution: Nguru Mts. and E. Usambara Mts., Tanzania.

Bradypodion fischeri excubitor (Barbour)

1911 *Chamaeleo tenuis excubitor* Barbour, Proc. biol. Soc., Washington, **24**: 219. – Terra typica: Mweru (= Meru) district, Mt. Kenya, Kenya. – Holotype (by orig. design.): MCZ R.782. – Syn.: fide Barbour & Loveridge 1928: 176.
1911 *Chamaeleon affinis embuensis* Lönnberg, K. Sv. Vetensk. Akad. Handl., Stockholm, **47**: 19; plate 2, fig. 3. – Terra typica: Embuboma, East Kenya. – Holotype (by implication): NRM 2960. – Syn.: fide Loveridge 1929: 91 and pers.observ.
1924 *Chamaeleon tenuis excurbitor* Loveridge (ex errore), J. East Afr. Uganda nat. Hist. Soc., **1924** (3): 16.
1928 *Chamaeleo fischeri excubitor* – Barbour & Loveridge, Mem. Mus. comp. Zool., **50**: 176.
1966 *Chamaeleo fischeri excubitor* – Mertens, Das Tierreich, **83**: 14.

Distribution: Mt. Kenya, Kenya.

Bradypodion fischeri multituberculatum (Nieden)

1913 *Chamaeleon fischeri multituberculatus* Nieden, S.Ber. Ges. naturf. Fr. Berlin, **1913**: 248; figs. 27–29; plate 15, fig. 7; plate 16, fig. 11. – Terra typica: Philippshof bei Wilhelmstal (= Magamba, near Lushoto), W. Usambara, Tanzania. – Type(s): not located.

1913 *Chamaeleon fischeri werneri* NIEDEN (non *Chamaeleon werneri* TORNIER 1899), S.Ber. Ges. naturf. Fr. Berlin, **1913**: 248; figs. 19–26, plate 15, fig. 6; plate 16, fig. 10. – Terra typica: Mlalo and Ambangula, W. Usambara, Tanzania. – Syntypes: ZMB 19266–19276 (11 specimens) and NMW 23382 (1 specimen). – Syn.: fide BARBOUR & LOVERIDGE 1928: 174.
1966 *Chamaeleo fischeri multituberculatus* – MERTENS, Das Tierreich, **83**: 14.

Distribution: W. Usambara Mts., Tanzania.

Bradypodion fischeri uluguruense (LOVERIDGE)

1957 *Chamaeleo fischeri uluguruensis* LOVERIDGE, Tanganyika Notes and Records, **43**: 3. – Terra typica: Kingokwa, Uluguru Mts., Tanzania. – Holotype (by orig. design): MCZ 54496.
1966 *Chamaeleo fischeri uluguruensis* – MERTENS, Das Tierreich, **83**: 15.

Distribution: Uluguru Mts., Tanzania.

Bradypodion mlanjense (BROADLEY)

1965 *Chamaeleo mlanjense* BROADLEY, Arnoldia, **1**: 1; figs. 1–2. – Terra typica: Ruo Gorge Forest on Mljanje Mountain, Malawi. – Holotype (by orig. design.): UM 4268.
1966 *Chamaeleo mlanjense* – MERTENS, Das Tierreich, **83**: 20.
1986 *Bradypodion mlanjense* – KLAVER & BÖHME, Bonn. zool. Monogr., **22**: 59.

Distribution: Known only from the type locality.

Bradypodion nemorale RAW

1978 *Bradypodion nemorale* RAW, Durban Mus. Novitates, **11**: 266; figs. 1–2. – Terra typica: Qudeni forest, Zululand, South Africa. – Holotype (by orig. design.): NM 1474.

Distribution: Qudeni and Nkandla forests, Zululand, South Africa.

Bradypodion oxyrhinum KLAVER & BÖHME

1988 *Bradypodion oxyrhinum* KLAVER & BÖHME, Bonn. zool. Beitr., **39**: 389; figs. 1, 5, 6B. – Terra typica: Mkoya in Ukami, Uluguru Mts., Tanzania. – Holotype (by orig. design.): ZFMK 46406.

Distribution: Uluguru Mts. and Uzungwe Mts., Tanzania.

Bradypodion pumilum (GMELIN)*

?1768 *Chamaeleo bonae spei* LAURENTI, Synops. Rept.: 46. – Terra typica: not given, but by implication: Cape of Good Hope, South Africa. – Type(s): not located; based on SEBA (1734) **I**; plate 83, fig. 5.
1789 *Lacerta pumila* GMELIN in LINNAEUS, Syst. Nat., 13nd ed., **1**: 1069. – Terra typica: Habitat ad caput bonae spei. – Type(s): not located.
1807 *C[hamaeleo] thermophilus* GRAVENHORST, Vergl. Übersicht des Linneischen etc.: 429. – Terra typica: not given. – Type(s): not located; based on SEBA (1734) **I**; plate 82, fig. 4 & plate 83, fig. 5.
1831 *Chameleo pumillus* GRAY (ex errore) in: GRIFFITH, Animal. Kindgom, **9** (Synops.): 53.
?1843 *Chamaeleon tunicatus* FITZINGER (nomen nudum), Syst. Rept., **1**: 43.
1865 *Lophosaura pumila* var. *fordii* Gray, Proc. zool. Soc. London, **1864**: 475. – Terra typica: South Africa. – Syntypes: BMNH 1946.8.21.8 and 1946.8.22.16 (2 specimens). – Syn.: fide BOULENGER 1887: 459.
1929 *Chamaeleon punilus* ANGEL (ex errore), Mém. Acad. malgache, Antananarivo, **9**: 56.

1966 *Chamaeleo pumilus* – MERTENS, Das Tierreich, **83**: 23.
1976 *Bradypodion pumilum* – RAW, Durban Mus. Novit., **11**: 141.

Distribution: S. Namibia, South Africa and S. Mozambique.

Bradypodion pumilum pumilum (GMELIN)

1820 *Chamaeleon margaritaceus* MERREM, Vers. Syst. Amph.: 162. – Terra typica: In Africa australi. – Type(s): PUM; presumed lost (H. O. von Hagen, in litt. 8.IX.1994). – Syn.: fide DUMÉRIL & BIBRON 1836: 217.
1911 *Chamaeleon pumilus* – WERNER, Das Tierreich, **27**: 22.
1959 *Chamaeleo pumilus pumilus* – HILLENIUS, Beaufortia, Amsterdam, **8**: 63.
1966 *Chamaeleo pumilus pumilus* – MERTENS, Das Tierreich, **83**: 23.

Distribution: South-western Cape Province, South Africa.

Bradypodion pumilum caffer (BOETTGER)

1889 *Chamaeleon caffer* BOETTGER, Ber. senckenb. naturf. Ges., **1889**: 292. – Terra typica: Pondoland, South Africa. – Holotype (by monotypy): not located; presumed lost (FITZSIMONS 1943: 162).
1911 *Chamaeleon caffer* – WERNER, Das Tierreich, **27**: 24.
1959 *Chamaeleo pumilus caffer* – HILLENIUS, Beaufortia, Amsterdam, **8**: 63.
1966 *Chamaeleo pumilus caffer* – MERTENS, Das Tierreich, **83**: 23.
1976 *Bradypodion caffrum* – RAW, Durban Mus. Novit., **11**: 141.

Distribution: Near Port St. Johns, Pondoland, eastern Cape Province and Natal, South Africa.

Bradypodion pumilum damaranum (BOULENGER)

1887 *Chamaeleon damaranus* BOULENGER, Cat. Lizards Brit. Mus., **3**: 460; plate 40, fig. 3. – Terra typica: Damaraland. – Terra typica designata (fide FITZSIMONS 1943: 166): Knysna, Cape Province. – Syntypes: BMNH 1946.9.4.78–79 (2 specimens).
1911 *Chamaeleon damaranus* – WERNER, Das Tierreich, **27**: 24.
1959 *Chamaeleo pumilus damaranus* – HILLENIUS, Beaufortia, Amsterdam, **8**: 63.
1966 *Chamaeleo pumilus damaranus* – MERTENS, Das Tierreich, **83**: 23.
1976 *Bradypodion damaranum* – RAW, Durban Mus. Novit., **11**: 149.

Distribution: Southern Cape Province, South Africa.

Bradypodion pumilum gutturale (SMITH)*

1849 *Chamaeleo gutturalis* SMITH, Illustr. Zool. S. Afr., Rept., App.: 3. – Terra typica: not given. – Type(s): not located; presumed lost (FITZSIMONS 1935/36: 270 & 1943: 168).
1911 *Chamaeleon gutturalis* – WERNER, Das Tierreich, **27**: 22.
1959 *Chamaeleo pumilus gutturalis* – HILLENIUS, Beaufortia, Amsterdam, **8**: 63.
1966 *Chamaeleo pumilus gutturalis* – MERTENS, Das Tierreich, **83**: 24.
1976 *Bradypodion gutturale* – RAW, Durban Mus. Novit., **11**: 149.

Distribution: Western Cape Province, South Africa.

Bradypodion pumilum karrooicum (METHUEN & HEWITT)

1915 *Lophosaura ventralis* var. *karrooica* METHUEN & HEWITT, Transact. roy. Soc. S. Afr., Cape Town, **4**: 103. – Terra typica: Beaufort West, Cape Province. – Holotype (by orig. design.): PEM R8942 (formerly AM 1789).

1935 *Lophosaura ventralis karooica* HEWITT (ex errore), Rec. Albany Mus., Grahamstown, **4**: 297.
1943 *Microsaura ventralis karroica* FITZSIMONS (ex errore), The lizards of South Africa: 168.
1959 *Chamaeleo pumilus karrooicus* – HILLENIUS, Beaufortia, Amsterdam, **8**: 63.
1966 *Chamaeleo pumilus karroicus* MERTENS (ex errore), Das Tierreich, **83**: 24.
1976 *Bradypodion karroicum* RAW (ex errore), Durban Mus. Novit., **11**: 149.

Distribution: Karroo, Cape Province, South Africa.

Bradypodion pumilum melanocephalum (GRAY)

1865 *Microsaura melanocephala* GRAY, Proc. zool. Soc. London, **1864**: 474; fig. – Terra typica: Port Natal, South Africa. – Holotype (by monotypy): BMNH 1946.8.21.7.
1911 *Chamaeleon melanocephalus* – WERNER, Das Tierreich, **27**: 22.
1935 *Lophosaura melanocephala kentanica* HEWITT, Rec. Albany Mus., Grahamstown, **4**: 303; plate 28, fig. 3. – Terra typica: Kentani, South Africa. – Syntypes: PEM, not registered (2 specimens, formerly AM). – Syn.: fide FITZSIMONS 1943: 160.
1959 *Chamaeleo pumilus melanocephalus* – HILLENIUS, Beaufortia, Amsterdam, **8**: 63.
1966 *Chamaeleo pumilus melanocephalus* – MERTENS, Das Tierreich, **83**: 24.
1966 *Lophosaura melanocephala kentamica* MERTENS (ex errore), Das Tierreich, **83**: 24.
1976 *Bradypodion melanocephalum* – RAW, Durban Mus. Novit., **11**: 149.

Distribution: Coastal regions of eastern Cape Province (Pondoland) and Natal, South Africa.

Bradypodion pumilum occidentale (HEWITT)

1935 *Lophosaura ventralis occidentalis* HEWITT, Rec. Albany Mus., Grahamstown, **4**: 297; plate 27, figs. 2–3. – Terra typica: Namaqualand, Cape Province, South Africa. – Holotype (by orig. design.): SAM 933.
1959 *Chamaeleo pumilus occidentalis* – HILLENIUS, Beaufortia, Amsterdam, **8**: 63.
1966 *Chamaeleo pumilus occidentalis* – MERTENS, Das Tierreich, **83**: 24.
1976 *Bradypodion occidentale* – RAW, Durban Mus. Novit., **11**: 149.

Distribution: Little Namaqualand, extending northwards along the coastal areas of Great Namaqualand, South Africa to Lüderitz, Namibia.

Bradypodion pumilum taeniabronchum (SMITH)

1831 *Chaemelio taeniabronchus* SMITH, S. Afr. quart. J., **5**: 17. – Terra typica: near Algoa Bay, Cape Province, South Africa. – Holotype (by monotypy): BMNH 1946.8.22.26.
1887 *Chamaeleon taeniobronchus* BOULENGER (nomen emendatum pro *Chaemelio taeniabronchus* SMITH 1831), Cat. Lizards Brit. Mus., **3**: 158; plate 40, fig. 1.
1911 *Chamaeleon taeniobronchus* – WERNER, Das Tierreich, **27**: 23.
1935 *Lophosaura taeniatrancha* POWER (ex errore), Proc. Zool. Soc. London, **1935**: 333.
1959 *Chamaeleo pumilus taeniabronchus* – HILLENIUS, Beaufortia, Amsterdam, **8**: 63.
1966 *Chamaeleo pumilus taeniabronchus* – MERTENS, Das Tierreich, **83**: 24.
1976 *Bradypodion taeniabronchum* – RAW, Durban Mus. Novit., **11**: 149.

Distribution: Near Algoa Bay, Cape Province, South Africa.

Bradypodion pumilum transvaalense (FITZSIMONS)

1930 *Chamaeleon transvaalensis* FITZSIMONS, Ann. Transvaal Mus., Pretoria, **14**: 37. – Terra typica: Haenertsburg, Transvaal, South Africa. – Holotype (by orig. design.): TM 2023.
1959 *Chamaeleo pumilus transvaalensis* – HILLENIUS, Beaufortia, Amsterdam, **8**: 63.
1966 *Chamaeleo pumilus transvaalensis* – MERTENS, Das Tierreich, **83**: 24.

1976 *Bradypodion transvaalense* – RAW, Durban Mus. Novit., **11**: 149.
?1977 *Chamaeleo natalensis* OBST (nomen nudum), Zool. Abh. Mus. Dresden, **34**: 168.

Distribution: Transvaal and Natal, South Africa.

Bradypodion pumilum ventrale (GRAY)

1845 *Chameleo ventralis* GRAY, Cat. Lizards Brit. Mus.: 268. – Terra typica: South Africa. – Syntypes: BMNH 1946.8.22.7–10 (4 specimens).
1865 *Chamaeleo pusillus* var.? GRAY (ex errore pro *pumilus*), Proc. zool. Soc. London, **1864**: 475.
1911 *Chamaeleon ventralis* – WERNER, Das Tierreich, **27**: 23.
1959 *Chamaeleo pumilus ventralis* – HILLENIUS, Beaufortia, Amsterdam, **8**: 63.
1966 *Chamaeleo pumilus ventralis* – MERTENS, Das Tierreich, **83**: 25.
1976 *Bradypodion ventrale* – RAW, Durban Mus. Novit., **11**: 149.

Distribution: Eastern Cape Province, South Africa.

Bradypodion setaroi RAW

1976 *Bradypodion setaroi* RAW, Durban Mus. Novitates, **11**: 152; fig. 4B; plate 1; map 1. – Terra typica: St. Lucia Estuary, Zululand, South Africa. – Holotype (by orig. design.): LRC 686.
1991 *Chamaeleo pumilus setaroi* – HOFMAN, MAXSON & ARNTZEN, Amphibia-Reptilia, **12**: 249.

Distribution: From St. Lucia to Lake Sibaya, South Africa.

Bradypodion spinosum (MATSCHIE)

1892 *Chamaeleon spinosus* MATSCHIE, S.Ber. Ges. naturf. Fr. Berlin, **1892**: 105. – Terra typica: Derema, Usambara Mts., Tanzania. – Holotype (by monotypy): ZMB 11350.
1911 *Chamaeleon spinosus* – WERNER, Das Tierreich, **27**: 38.
1966 *Chamaeleo spinosus* – MERTENS, Das Tierreich, **83**: 27.
1986 *Bradypodion spinosum* – KLAVER & BÖHME, Bonn. zool. Monogr., **22**: 59.

Distribution: Usambara Mts., Tanzania.

Bradypodion tavetanum (STEINDACHNER)

1891 *Chamaeleon tavetanus* STEINDACHNER, Anz. Akad. Wiss. Wien, math.-naturw. Kl., **28**: 142. – Terra typica: Taveta-Wald, Kenya. – Syntypes: NMW 7919: 1,2 (2 specimens).
1891 *Chamaeleon tavetensis* STEINDACHNER (nomen substitutum pro *Chamaeleon tavetanus* STEINDACHNER 1891), S.Ber. Akad. Wiss. Wien (Abt. 1), **100**: 312; plate 1, figs. 3–3A.
1891 *Chamaeleo abbotti* STEJNEGER, Proc. U.S.nation. Mus., Washington, **14**: 353; fig. – Terra typica: Kilimanjaro (4500 ft. alt.), Tanganyika (= Tanzania). – Holotype (by orig. design.): USNM 16744. – Syn.: fide BOULENGER 1892: 73.
1893 *Chamaeleo taitensis* STEJNEGER (ex errore), Proc. U.S. nation. Mus., Washington, **16**: 724.
1911 *Chamaeleon tavetensis* – WERNER, Das Tierreich, **27**: 29; fig. 8.
1957 *Chamaeleo fischeri tavetanus* – LOVERIDGE, Bull. Mus. comp. Zool., **117**: 203.
1966 *Chamaeleo fischeri tavetanus* – MERTENS, Das Tierreich, **83**: 14.
1991 *Bradypodion tavetanum* – BROADLEY & HOWELL, Syntarsus, Bulawayo, **1**: 12.

Distribution: Teita Mts., Kenya and west to Kilimanjaro and Arusha (fide LOVERIDGE 1959: 32) and south to the N. Pare Mts. (fide MERTENS 1955: 54), Tanzania.

Bradypodion tenue (MATSCHIE)

1892 *Chamaeleon tenuis* MATSCHIE, S.Ber. Ges. naturf. Fr. Berlin, **1892**: 106. – Terra typica: Derema, Usambara Mts. Tanzania. – Lectotype: ZMB 11348.1 (fide KLAVER & BÖHME 1988: 389).
1911 *Chamaeleon tenuis* – WERNER, Das Tierreich, **27**: 38.
1911 *Chamaeleo tenuis tenuis* – BARBOUR, Proc. biol. Soc. Washington, **24**: 219.
1966 *Chamaeleo tenuis* – MERTENS, Das Tierreich, **83**: 28.
1986 *Bradypodion tenue* – KLAVER & BÖHME, Bonn. zool. Monogr., **22**: 59.

Distribution: Shimba Hills, Kenya and Usambara Mts., Tanzania (fide KLAVER & BÖHME 1988).

Bradypodion thamnobates RAW

1976 *Bradypodion thamnobates* RAW, Durban Mus. Novitates, **11**: 154; fig. 4C, plate 1, map 1. – Terra typica: Nottingham Road, Natal, South Africa. – Holotype (by orig. design.): LRC 825.
1991 *Chamaeleo pumilus thamnobates* – HOFMAN, MAXSON & ARNTZEN, Amphibia-Reptilia, **12**: 250.

Distribution: Natal Midlands from Mooi River to Bulwer and Dargle (1000–1400 m alt.), South Africa.

Bradypodion uthmoelleri (MÜLLER)

1938 *Chamaeleo uthmölleri* MÜLLER, Zool. Anz., **122**: 20; figs. 1–2. – Terra typica: Hanang (Gurui; 2300 m alt.), Tanzania. – Holotype (by orig. design.): ZSM 1/1948.
1957 *Chamaeleo fischeri uthmölleri* – LOVERIDGE, Bull. Mus. comp. Zool., **117**: 203.
1966 *Chamaeleo fischeri uthmoelleri* – MERTENS, Das Tierreich, **83**: 15.
1986 *Bradypodion uthmoelleri* – KLAVER & BÖHME, Bonn. zool. Monogr., **22**: 59.

Distribution: Mt. Hanang, Ngorongoro Crater and Oldeani, Tanzania (fide BÖHME & KLAVER 1990: 264).

Bradypodion xenorhinum (BOULENGER)

1901 *Chamaeleon xenorhinus* BOULENGER, Proc. zool. soc. London, **1901**: 135; plate 12. – Terra typica: Mount Ruwenzori (6000 ft. alt.), Uganda. – Syntypes: BMNH 1946.8.21.64–65 (2 specimens).
1911 *Chamaeleon xenorhinus* – WERNER, Das Tierreich, **27**: 29.
1966 *Chamaeleo xenorhinus* – MERTENS, Das Tierreich, **83**: 29.
1986 *Bradypodion xenorhinum* – KLAVER & BÖHME, Bonn. zool. Monogr., **22**: 59.

Distribution: Ruwenzori Mts., Uganda and Zaire.

Genus Chamaeleo LAURENTI

1763 *Chamaeleon* GRONOVIUS (nomen rejectum, fide Opinion 89), Zoophyl., **1**: 12.
1768 *Chamaeleo* LAURENTI, Synops. Rept.: 45. – Species typica (by tautonomy, fide: STEJNEGER 1936: 136): *Chamaeleo parisiensium* LAURENTI (= *Lacerta chamaeleon* LINNAEUS 1758).
1789 *Chameleo* BONNATERRE (ex errore), Tabl. encyclop. méth., Erpét.: 31.
1803 *Cameleo* BOSC (ex errore), Nouv. Dict., **4**: 157.
1806 *Camaeleo* DUMÉRIL (ex errore), Zool. analyt.: 82, 333.
1822 *Chameleon* FLEMING (nomen emendatum), Phil. Zool., **2**: 272.
1825 *Chamelion* GRAY (ex errore), Ann. Phil. Lond., (n.s.) **10**: 200.
1831 *Chaemelio* SMITH (ex errore), S.Afr. quart. J., **5**: 17.
1836 *Camaeleon* DUMÉRIL & BIBRON (ex errore), Erpétol. gén., **3**: 216.

1843 *Chaemaeleon* FITZINGER (ex errore), Syst. Rept.: 41.
1868 *Camaleo* ARADAS (ex errore), Atti Soc. ital. Sci. nat., **11**: 440.
1870 *Chamelio* BLANFORD (ex errore), Obs. Abyssinia: 445.
1885 *Chamaleo* BOETTGER (ex errore), Ber. offenb. Ver. Naturk., **24/25**: 175; footnote.
1887 *Chamaelleo* BOCAGE (ex errore), J. Acad. Sci., Lisboa, **12**: 85.
1891 *Chamaelon* STEINDACHNER (ex errore), S.Ber. Akad. Wiss., math.-naturw. Kl., Wien, **100**: 307.
1893 *Chamaelon* MATSCHIE (ex errore), S.Ber. Ges. naturf. Fr. Berlin, **1893**: 27.
1903 *Chamaello* BARBOUR (ex errore), Proc. biol. Soc. Washington, **16**: 61.
1931 *Chamadeon* PARKER (ex errore), Ann. Mag. nat. Hist., London, **8**: 516.
1957 *Chaelio* LOVERIDGE (ex errore), Bull. Mus. comp. Zool., **117**: 196.
1981 *Chamaelo* BÖHME & KLAVER (ex errore), Amphibia-Reptilia, **1**: 317.
1986 *Chaemaleo* HILLENIUS (ex errore), Bijdr. Dierk, Amsterdam, **56**: 33.

Distribution: Southern Europe, Near East, Arabian peninsula, Africa, S.E. Pakistan, India and Sri Lanka.

Subgenus Chamaeleo LAURENTI*

1768 *Chamaeleo* LAURENTI, Synops. Rept.: 45. – Species typica (by tautonomy, fide STEJNEGER 1936: 136): *Chamaeleo parisiensium* LAURENTI (= *Lacerta chamaeleon* LINNAEUS 1758).
1843 *Chamaeleon* FITZINGER (non *Chamaeleon* SCUDDER 1882 = Mollusca), Syst. Rept.: 15. – Species typica (by original designation): *Chamaeleon africanus* CUVIER (= LAURENTI 1768).
1865 *Phumanola* Gray, Proc. zool. Soc. London, **1864**: 467, 474. – Species typica (by monotypy): *Chamaeleo namaquensis* SMITH 1831.
1865 *Calyptrosaura* Gray, Proc. zool. Soc. London, **1864**: 468. – Species typica (by subsequent designation: LOVERIDGE 1957a: 197): *Chamaeleo calyptratus* DUMÉRIL & BIBRON 1851.
1865 *Erizia* GRAY, Proc. zool. Soc. London, **1864**: 471. – Species typica (by subsequent designation: LOVERIDGE 1957a: 197): *Chamaeleo senegalensis* DAUDIN 1802.
1865 *Dilepis* GRAY (non *Dilepis* WEINLAND 1858 = Cestoda), Proc. zool. soc. London, **1864**: 472. – Species typica (by tautonomy): *Chamaeleon dilepis* LEACH 1819.

Distribution: Southern Europe, Near East, Arabian peninsula, Africa. S.E. Pakistan, India and Sri Lanka.

Chamaeleo (Chamaeleo) **africanus** LAURENTI*

1768 *Chamaeleo africanus* LAURENTI, Synops. Rept.: 46. – Terra typica: not given. – Terra typica: restricta: Africa (fide MERTENS 1966: 6). – Type(s): not located; based on SEBA (1734) **I**; plate 43, fig. 4 (43 is a misprint for 83).
?1768 *Chamaeleo mexicanus* LAURENTI, Synops. Rept.: 45. – Terra typica: not given. – Type(s): not located; based on SEBA (1734) **I**; plate 82, fig. 1. – Syn.: fide FLOWER 1933: 782.
?1768 *Chamaeleo candidus* LAURENTI, Synops. Rept.: 46. – Terra typica: not given. – Type(s): not located; specimen in Museo Turriano. – Syn.: fide MERTENS 1966: 6.
1789 *Chameleo aegiptius* BONNATERRE, Tabl. encyclop., Erpétol.: 31; plate 7, fig. 2. – Terra typica: not given. – Terra typica restricta: Egypt. – Type(s): not located. – Syn.: fide MERTENS 1966: 6.
1820 *Chamaeleo calcaratus* MERREM, Vers. Syst. Amph.: 162. – Terra typica: Africa. – Type(s): PUM; presumed lost (H.O. von Hagen, in litt. 8.XI.1994). – Syn.: fide FLOWER 1933: 783.
1868 *Chamaeleo basiliscus* COPE, Proc. Acad. nat. Sci., Philadelphia, **20**: 316. – Terra typica: Korosko, El Der district, Province Assam, Upper Egypt. – Holotype (by monotypy): MCZ 5766. Syn.: fide FLOWER 1933: 783.
1911 *Chamaeleon basiliscus* – WERNER, Das Tierreich, **27**: 11; fig. 2.
1917 *Chamaeleon basilicus* STERNFELD (ex errore), Ergebn. zweit. dtsch. Zentr.-Afr.-Exped., 1910–1911, **1**: 409.

1966 *Chamaeleo africanus* – MERTENS, Das Tierreich, **83**: 17.
1986 *Chamaeleo (Chamaeleo) africanus* – KLAVER & BÖHME, Bonn. zool. Monogr., **22**: 59.

Distribution: Subsaharan Africa, from Mali, Niger, Nigeria, Cameroon and Gabon to Egypt, Sudan and Eritrea.

Chamaeleo (Chamaeleo) **anchietae** BOCAGE*

1872 *Chamaeleo anchietae* BOCAGE, J. Sci., Lisboa, **4**: 72; fig. – Terra typica: Huila, Mossamedes, Angola. – Syntypes: MBL (5 specimens); lost by fire (fide SALDANHA 1978).
1911 *Chamaeleon anchietae* – WERNER, Das Tierreich, **27**: 16; fig. 4.
1950 *Chamaeleo (Apola) vinckei* LAURENT, Rev. zool. bot., Bruxelles, **43**: 349. – Terra typica: Station d'étude des Kundelungu (1750 m alt.), Katanga, Zaire. – Holotype (by orig. design.): MRAC 26.261. – Syn.: fide LOVERIDGE 1957a: 198.
1952 *Chamaeleo anchietae anchietae* – LAURENT, Rev. zool. bot., Bruxelles **46**: 19.
1952 *Chamaeleo anchietae mertensi* LAURENT, Rev. zool. bot., Bruxelles, **46**: 18. – Terra typica: Muneshele (1400 m alt.), Terr. de Fizi, Kivu, Zaire. – Holotype (by orig. design.): MRAC 26.270. – Syn.: fide HILLENIUS 1959: 51.
1952 *Chamaeleo anchietae marunguensis* LAURENT, Rev. zool. bot., Bruxelles, **46**: 19. – Terra typica: Kipiri (2000 m alt.), Plateau des Marungu, Terr. de Baudouinville, W. of Lake Tanganyika, Zaire. – Holotype (by orig. design.): MRAC 26.273. – Syn.: fide HILLENIUS 1959: 51.
1952 *Chamaeleo anchietae vinckei* – LAURENT, Rev. zool. bot. Bruxelles, **46**: 19.
1957 *Chamaeleo senegalensis anchietae* – LOVERIDGE, Bull. Mus. comp. Zool., **117**: 198.
1966 *Chamaeleo anchietae* – MERTENS, Das Tierreich, **83**: 7.
1966 *Chamaeleo anchietae anchietae* – MERTENS, Das Tierreich, **83**: 7.
1966 *Chamaeleo anchietae marunguensis* – MERTENS, Das Tierreich, **83**: 7.
1966 *Chamaeleo anchietae mertensi* – MERTENS, Das Tierreich, **83**: 7.
1966 *Chamaeleo anchietae vinckei* – MERTENS, Das Tierreich, **83**: 7.
1986 *Chamaeleo (Chamaeleo) anchietae* – KLAVER & BÖHME, Bonn. zool. Monogr., **22**: 59.

Distribution: Angola, S.E. Zaire, Rwanda and S. Tanzania.

Chamaeleo (Chamaeleo) **arabicus** MATSCHIE

1893 *Chamaelon arabicum* MATSCHIE, S.Ber. Ges. naturf. Fr. Berlin, **1893**: 27. – Terra typica: Gärten bei Lahadsh (= Oasis Lahej near Aden), Yemen. – Syntypes: ZMB 11135 & 38572 (2 specimens).
1911 *Chamaeleon calcarifer* – WERNER (partim) Das Tierreich, **27**: 10.
1966 *Chamaeleo chamaeleon arabicus* – HILLENIUS, Beaufortia, Amsterdam, **13**: 104.
1966 *Chamaeleo chamaeleon calcarifer* – MERTENS (partim), Das Tierreich, **83**: 10.
1986 *Chamaeleo (Chamaeleo) arabicus* – KLAVER & BÖHME, Bonn. zool. Monogr., **22**: 18, 59.

Distribution: Oman and Yemen (fide MEERMAN & BOOMSMA 1987: 10).

Chamaeleo (Chamaeleo) **calcaricarens** BÖHME

1985 *Chamaeleo (africanus) calcaricarens* BÖHME, Proc. internat. Symp. Afr. Vertebr., Bonn: 475. – Terra typica: Gewani, Ethiopian Danakil, Ethiopia. – Holotype (by orig. design.): BMNH 1961.1774.
1986 *Chamaeleo (Chamaeleo) calcaricarens* – KLAVER & BÖHME, Bonn. zool. Monogr., **22**: 59.

Distribution: Ethiopia, Eritrea, Djibouti and Somalia.

Chamaeleo (Chamaeleo) **calyptratus** DUMÉRIL & BIBRON*

1851 *Chamaeleo calyptratus* DUMÉRIL & BIBRON in: DUMÉRIL & DUMÉRIL, Cat. méth. Coll. Rept.: 31. – Terra typica: région du Nil, Africa. – Syntypes: MHNP 6522, 6633–34 (3 specimens).

1911 *Chamaeleon calyptratus* – WERNER, Das Tierreich, **27**: 17.
1966 *Chamaeleo calyptratus* – MERTENS, Das Tierreich, **83**: 9.
1986 *Chamaeleo (Chamaeleo) calyptratrus* – KLAVER & BÖHME, Bonn. zool. Monogr., **22**: 59.

Distribution: S.W. Saudi Arabia and Yemen.

Chamaeleo (Chamaeleo) calyptratus calyptratus DUMÉRIL & BIBRON

1985 *Chamaeleo calyptratus calyptratus* – HILLENIUS & GASPERETTI, Fauna of Saudi Arabia, **6**: 525; figs. 9–10.

Distribution: Yemen.

Chamaeleo (Chamaeleo) calyptratus calcarifer PETERS*

1869 *Chamaeleo calcaratus* PETERS (non *Chamaeleo calcaratus* MERREM 1820), M.Ber. Akad. Wiss. Berlin, **1869**: 445. – Terra typica: Bembatuka, west coast of Madagascar. – Terra typica designata (fide MERTENS 1966: 10): southern Arabian peninsula. – Holotype (by monotypy): ZMB 316.
1870 *Chamaeleo calcarifer* PETERS (nomen novum pro *Chamaeleo calcaratus* PETERS 1869), M.Ber. Akad. Wiss. Berlin, **1870**: 110; footnote.
1911 *Chamaeleon calcarifer* – WERNER (partim), Das Tierreich, **27**: 10.
1959 *Chamaeleo chamaeleon calcarifer* – HILLENIUS, Beaufortia, Amsterdam, **8**: 45.
1966 *Chamaeleo chamaeleon calcarifer* – MERTENS (partim), Das Tierreich, **83**: 10.
1985 *Chamaeleo calyptratus calcarifer* – HILLENIUS & GASPERETTI, Fauna of Saudi Arabia, **6**: 520; plate 3–4, figs. 11 & 13.

Distribution: S.W. Saudi Arabia and possibly Yemen (fide FRITZ & SCHÜTTE 1987: 23).

Chamaeleo (Chamaeleo) **chamaeleon** (LINNAEUS)*

1758 *Lacerta chamaeleon* LINNAEUS, Syst. Nat., 10th ed., **1**: 204. – Terra typica: Habitat in Africa, Asia. – Terra typica restricta (fide MERTENS & MÜLLER 1928: 25): Nordafrika. – Syntypes: NRM (2 specimens; fide ANDERSSON 1900: 13).
1795 *Chamaeleon mntabilis* Meyer (ex errore, nomen substitutum pro *Lacerta chamaeleon* LINNAEUS 1758), Synops. Rept.: 27.
1802 *Chamaeleon vulgaris* DAUDIN (nomen substitutum pro *Lacerta chamaeleon* LINNAEUS 1758), Hist. Nat. Rept., **4**: 181.
?1866 *Chamaeleon fasciatus* SMITH, Proc. roy. phys. Soc., Edinburgh, **3**: 306. – Terra typica: Old Calabar. – Type(s): not located. – Syn.: fide BOULENGER 1887: 443.
1911 *Chamaeleon chamaeleon* – WERNER, Das Tierreich, **27**: 10.
1966 *Chamaeleo chamaeleon* – MERTENS, Das Tierreich, **83**: 9.
1986 *Chamaeleo (Chamaeleo) chamaeleon* – KLAVER & BÖHME, Bonn. zool. Monogr., **22**: 59.

Distribution: Southern Europa, North Africa, Near East and Arabian peninsula.

Chamaeleo (Chamaeleo) chamaeleon chamaeleon (LINNAEUS)*

1768 *Chamaeleo parisiensium* LAURENTI, Synops. Rept.: 45. – Terra typica: India orientali. – Type(s): not located; based on SEBA (1734) **I**; plate 82, fig. 2, 4 & 5. – Syn.: fide STEJNEGER 1936: 136.
1820 *Chamaeleon carinatus* MERREM (nomen substitutum pro *Chamaeleo parisiensium* Laurenti 1768), Vers. Syst. Amph.: 162.
1836 *Cameleo siculus* DUMÉRIL & BIBRON (nomen nudum), Erpétol. gén., **3**: 205.
1843 *Chamaeleon hispanicus* FITZINGER (nomen nudum), Syst. Rept., **1**: 42.

1843 *Chamaeleon rimulosus* FITZINGER, (nomen nudum), Syst. Rept., **1**: 42.
1845 *Chameleo parisiensis* GRAY (nomen substitutum pro *Chamaeleo parisiensium* LAURENTI 1768), Cat. Liz. Brit. Mus.: 265.
1862 *Chamaeleo cinereus* STRAUCH, Mém. Acad. Sci., St. Pétersbourg, **6**: 21. – Terra typica: various localities in Algeria. – Terra typica restricta (fide MERTENS & WERMUTH 1960: 85; non CRESPO 1972: 475): Algeria. – Type(s): ?ZISP; not located. – Syn.: fide BOULENGER 1887: 443.
1865 *Chamaeleon auratus* GRAY, Proc. zool. Soc. London, **1864**: 469. – Terra typica: Arabia. – Holotype (by monotypy): BMNH 1946.8.21.11. – Syn.: fide BOULENGER 1887: 443.
?1868 *Camaleo mitratus* ARADAS (nomen nudum), Atti Soc. ital. Sci. nat., Milano, **11**: 443.
1880 *Ch[amaeleon] parisientium* BOSCÁ (ex errore), Bull. Soc. zool. France, **5**: 282.
1880 *Chamaeleo vulgaris* var. *recticrista* BOETTGER, Ber. senckenb. naturf. Ges., **1879/1880**: 198. – Terra typica: Haifa, Jerusalem, Beirut and Cyprus. – Terra typica restricta (fide MERTENS 1922: 177): Jerusalem. – Lectotypus: SMF 16151 (fide MERTENS 1967: 65). – Syn.: fide HILLENIUS 1978: 54.
1887 *Chamaeleo saharicus* MÜLLER, Verh. naturf. Ges., Basel, **8**: 295. – Terra typica: Boussaada, Algeria. – Holotype (by monotypy): NMB 4884. – Syn.: fide WERNER, 1902: 328.
1911 *Chamaeleon chamaeleon chamaeleon* – WERNER, Das Tierreich, **27**: 10.
1911 *Chamaeleon chamaeleon saharicus* – WERNER, Das Tierreich, **27**: 10.
1966 *Chamaeleo chamaeleon chamaeleon* – MERTENS, Das Tierreich, **83**: 10.
1966 *Chamaeleo parisiensicum* MERTENS (ex errore), Das Tierreich, **83**: 36.
1966 *Cameleo siculus* MERTENS (ex errore), Das Tierreich, **83**: 10.
1966 *Chamaeleo chamaeleon recticrista* – MERTENS, Das Tierreich, **83**: 11.

Distribution: Southern Portugal, southern Spain, Sicily, Malta, S. Greece, Crete, Samos, Chios, Cyprus, S. Turkey, Libanon, Syria, Israel, Jordan, Egypt, Libya, Tunesia, Algeria, Morocco and Western Sahara; introduced in the Canary Islands (Lanzarote) (fide WERNER 1902: 329).

Chamaeleo (Chamaeleo) chamaeleon musae STEINDACHNER

1901 *Chamaeleon vulgaris* var. *musae* STEINDACHNER, Denkschr. Akad. Wiss. Wien, math.-naturw. Kl., **69**: 331. – Terra typica: Ain Musa, Sinai peninsula, Egypt. – Syntypes: NMW 18163: 1–5 (5 specimens).
1911 *Chamaeleon chamaeleon musae* – WERNER, Das Tierreich, **27**: 10.
1966 *Chamaeleo chamaeleon musae* – MERTENS, Das Tierreich, **83**: 10.

Distribution: Sinai peninsula, Egypt.

Chamaeleo (Chamaeleo) chamaeleon orientalis PARKER

1938 *Chamaeleo chamaeleon orientalis* PARKER, Ann. Mag. nat. Hist., London, **1**: 486. – Terra typica: Taif (5500 ft. alt.), S. of Hijaz, Saudi Arabia. – Holotype (by orig.design.): BMNH 1935.10.20.15.
1966 *Chamaeleo chamaeleon orientalis* – MERTENS, Das Tierreich, **83**: 11.

Distribution: S.W. Saudi Arabia and Yemen.

Chamaeleo (Chamaeleo) **dilepis** LEACH*

1819 *Chamaeleon dilepis* LEACH in BODWICH, Mission from Cape Coast Castle to Ashantee, App. IV: 493. – Terra typica: French Congo (= Congo). – Holotype (by monotypy): BMNH 1946.8.13.75.
1820 *Chamaeleo planiceps* MERREM, Vers. Syst. Amph.: 162. – Terra typica: Africa. – Type(s): PUM; presumed lost (H.O. von Hagen, in litt. 8.XI.1994). – Syn.: fide DUMÉRIL & BIBRON 1836: 225.
1837 *Chameleon dilepas* STUTCHBURY (ex errore), Trans. Linn. Soc., London, **17**: 362.

1838 *Chamaeleon dilepas* Martin (ex errore), Proc. zool. Soc. London, **1838**: 64.
1911 *Chamaeleon dilepis* – WERNER (partim), Das Tierreich, **27**: 13.
1966 *Chamaeleo dilepis* – MERTENS, Das Tierreich, **83**: 12.
1986 *Chamaeleo (Chamaeleo) dilepis* – KLAVER & BÖHME, Bonn. zool. Monogr., **22**: 59.

Distribution: Tropical and southern Africa, except for the countries west of Nigeria.

Chamaeleo (Chamaeleo) dilepis dilepis LEACH

1820 *Chamaeleo bilobus* KUHL, Beitr. Zool. vergl. Anat., **1**: 104. – Terra typica: Französisch-Kongo (= Congo). – Type(s): BMNH, not registered as type(s). – Syn.: fide DUMÉRIL & BIBRON 1836: 225.
1865 *Chamaeleon petersii* var. *kirkii* GRAY, Proc. zool. Soc. London, **1864**: 470. – Terra typica: Eastern Africa. – Holotype (by monotypy): BMNH 1946.8.22.33. – Syn.: fide BOULENGER 1887: 450.
1866 *Chamaeleo capellii* BOCAGE, J. Sci., Lisboa **1**: 42, 59. – Terra typica: Benguela, Angola. – Holotype (by monotypy): MBL, lost by fire (fide SALDANHA 1978). – Syn.: fide BOULENGER 1887: 450.
1903 *Chamaeleo angusticoronatus* BARBOUR, Proc. biol. Soc. Washington, **16**: 61. – Terra typica: Zanzibar, Tanzania. – Holotype (by orig. design.): MCZ 6712. – Syn.: fide LOVERIDGE 1929: 87.
1911 *Chamaeleon dilepis dilepis* – WERNER, Das Tierreich, **27**: 13.
1911 *Chamaeleon angusticoronatus* – WERNER, Das Tierreich, **27**: 14.
1966 *Chamaeleo dilepis dilepis* – MERTENS, Das Tierreich, **83**: 12.

Distribution: Gabon, Equatorial Guinea, Congo, Zaire, Angola, Somalia, Uganda, Kenya, Tanzania, Zambia, Malawi, Mozambique, Zimbabwe, Namibia, South Africa (Natal, Transvaal) and Botswana.

Chamaeleo (Chamaeleo) dilepis idjwiensis LOVERIDGE

1942 *Chamaeleo dilepis idjwiensis* LOVERIDGE, Bull. Mus. comp. Zool., **91**: 363; plate 5, fig. 2. – Terra typica: Upper Mulinga River, Idjwi Island, Lake Kivu, Belgian Congo (= Zaire). – Holotype (by orig. design.): MCZ 47155.
1966 *Chamaeleo dilepis idjwiensis* – MERTENS, Das Tierreich, **83**: 12.

Distribution: Burundi, Rwanda, Uganda and E. Zaire.

Chamaeleo (Chamaeleo) dilepis isabellinus GÜNTHER

1893 *Chamaeleon isabellinus* Günther, Proc. zool. Soc. London, **1892**: 556; plate 33, fig. 2. – Terra typica: Shiré Highlands south of Laka Nyassa, principally upon Mount Zomba and Mount Milanje, Nyassaland (= Malawi). – Holotype (by monotypy): BMNH 1946.8.21.5.
1896 *Chamaeleon isabelinus* BOCAGE (ex errore), J. Acad. Sci., Lisboa, **4**: 103.
1911 *Chamaeleon dilepis isabellinus* – WERNER (partim), Das Tierreich, **27**: 14.
1966 *Chamaeleo dilepis isabellinus* – MERTENS, Das Tierreich, **83**: 13.

Distribution: Shire Highlands, Malawi.

Chamaeleo (Chamaeleo) dilepis martensi MERTENS

1964 *Chamaeleo dilepis martensi* MERTENS, Senckenb. biol., **45**: 113: fig. 1. – Terra typica: Insel Pemba, Tanzania. – Holotype (by orig. design.): SMF 16144.
1966 *Chamaeleo dilepis martensi* – MERTENS, Das Tierreich, **83**: 13.

Distribution: Only known from the type locality.

Chamaeleo (Chamaeleo) dilepis petersii GRAY

1865 *Chamaeleon petersii* GRAY, Proc. zool. soc. London, **1864**: 470; fig. – Terra typica: Mozambique. – Syntypes: BMNH 1946.8.3.67–69 and 1947.3.6.54 (4 specimens).
1911 *Chamaeleon dilepis* – WERNER (partim), Das Tierreich, **27**: 13.
1953 *Chamaeleo dilepis petersii* – LOVERIDGE, Bull. Mus. comp. Zool., **110**: 186.
1966 *Chamaeleo dilepis petersii* – MERTENS, Das Tierreich, **83**: 13.

Distribution: Mozambique and Malawi.

Chamaeleo (Chamaeleo) dilepis roperi BOULENGER

1890 *Chamaeleon roperi* BOULENGER, Proc. zool. Soc. London, **1890**: 85; plate 8, fig. 4. – Terra typica: Kilifi, north of Mombasa, Kenya. – Syntypes: BMNH 1946.8.22.1–2 (2 specimens).
1902 *Chamaeleon dilepis* var. *roperi* – WERNER, Zool. Jb., Syst., **15**: 343.
1911 *Chamaeleon dilepis roperi* – WERNER, Das Tierreich, **27**: 13.
1966 *Chamaeleo dilepis roperi* – MERTENS, Das Tierreich, **83**: 13.

Distribution: Gabon, Angola, E. Kenya, Uganda, N.E. Tanzania and Zaire.

Chamaeleo (Chamaeleo) dilepis ruspolii BOETTGER*

1893 *Chamaeleon ruspolii* BOETTGER, Zool. Anz., **16**: 116. – Terra typica: Ogaden, Somaliland. – Lectotypus (hoc loco): SMF 16300.
1911 *Chamaeleon dilepis isabellinus* – WERNER (partim), Das Tierreich, **27**: 14.
1942 *Chamaeleo dilepis ruspolii* – PARKER, Bull. Mus. comp. Zool., **91**: 79.
1966 *Chamaeleo dilepis ruspolii* – MERTENS, Das Tierreich, **82**: 13.

Distribution: Ethiopia (fide MERTENS 1967: 65).

Chamaeleo (Chamaeleo) **gracilis** HALLOWELL

1842 *Chamaeleo gracilis* HALLOWELL, J. Acad. nat. Sci., Philadelphia, **8**: 324; plate 18. – Terra typica: Liberia, West Africa. – Holotype (by monotypy): ANSP 7325.
1966 *Chamaeleo gracilis* – MERTENS, Das Tierreich, **83**: 16.
1986 *Chamaeleo (Chamaeleo) gracilis* – KLAVER & BÖHME, Bonn. zool. Monogr., **22**: 59.

Distribution: Tropical Africa; from Senegal to Angola and from Sudan to N. Tanzania.

Chamaeleo (Chamaeleo) gracilis gracilis HALLOWELL

1857 *Chamaeleo granulosus* HALLOWELL, Proc. Acad. nat. Sci., Philadelphia, **1856**: 147. – Terra typica: West Africa. – Holotype (by monotypy): ANSP 7322. – Syn. fide BOULENGER 1887: 448.
?1857 *Chamaeleo burchelli* HALLOWELL, Proc. Acad. nat. Sci., Philadelphia, **1856**: 147. – Terra typica: Fernando Poo (= Bioko, Equatorial Guinea). – Holotype (by monotypy): ANSP 7329. – Syn.: fide BOULENGER 1887: 448.
1885 *Chamaleo (Chamaleo) simoni* BOETTGER, Ber. offenb. Ver. Naturk., **24/25**: footnote. – Terra typica: Berge zwischen Goldküste und Ashanti-Land, Goldküste (= Ghana). – Lectotypus: SMF 16218 (fide MERTENS 1967: 64). – Syn.: fide BOULENGER 1887: 448.
1911 *Chamaeleon gracilis* – WERNER, Das Tierreich, **27**: 12.
1929 *Chamaeleo gracilis gracilis* – LOVERIDGE, U.S. nation. Mus. Bull., **151**: 84.
1966 *Chamaeleo gracilis gracilis* – MERTENS, Das Tierreich, **83**: 16.

Distribution: Senegal, Gambia, Guinea, Sierra Leone, Liberia, Ivory Coast, Ghana, Nigeria, Cameroon, Gabon, Central African Republic, Congo, Zaire, Tanzania, Kenya, Uganda, Somalia, Ethiopia and Sudan.

Chamaeleo (Chamaeleo) gracilis etiennei SCHMIDT

1919 *Chamaeleon etiennei* SCHMIDT, Bull. Amer. Mus. nat. Hist., **39**: 574. – Terra typica: Banana, Zaire. – Holotype (by orig. design.): AMNH 11370.
1929 *Chamaeleo gracilis etiennei* – LOVERIDGE, U.S. nation. Mus. Bull., **151**: 84.
1966 *Chamaeleo gracilis etiennei* – MERTENS, Das Tierreich, **83**: 17.

Distribution: Gabon, Congo, Zaire and Angola.

Chamaeleo (Chamaeleo) **laevigatus** GRAY*

1863 *Chameleo laevigatus* GRAY, Proc. zool. Soc. London, **1863**: 95. – Terra typica: Khartoom, Sudan; later corrected in: 500 miles south of Khartoom (fide BOULENGER 1887: 448). – Syntypes: BMNH 1946.8.22.5–6 (2 specimens).
1887 *Chamaeleon sphaeropholis* REICHENOW, Zool. Anz., **10**: 370. – Terra typica: Kagehi (= Kagera), westlich des Victoria-Sees, Tanganyika (= Tanzania). – Holotype (by monotypy): ZMB 10745. – Syn.: fide WERNER 1911: 18.
1888 *Chamaeleon senegalensis laevigatus* – GÜNTHER, Proc. zool. Soc. London, **1888**: 50.
1902 *Chamaeleon lavigatus* BOULENGER (ex errore), Uganda Protectorate, **1**: 446.
1911 *Chamaeleon laevigatus* – WERNER, Das Tierreich, **27**: 18.
1937 *Chamaeleo laeviceps* PASQUAL (nomen nudum), Nig. Field, **6**: 34.
1951 *Chamaeleon senegalensis* var. *tibatiensis* MONARD, Mém. Intern. fr. Afr. noire (Sci.nat.), Dakar, **1**: 141. – Terra typica: Tibati, Cameroon. – Lectotype (hoc loco): MHNC 91.063. – Syn.: fide PERRET & MERTENS 1957: 581.
1957 *Chamaeleo senegalensis tibiatensis* PERRET (ex errore), Rev. suisse Zool., **64**: 85.
1966 *Chamaeleo senegalensis laevigatus* – MERTENS, Das Tierreich, **83**: 27.
1986 *Chamaeleo (Chamaeleo) laevigatus* – KLAVER & BÖHME, Bonn. zool. Monogr., **22**: 21, 59.

Distribution: Cameroon, Central African Republic, Zaire, Rwanda, Burundi, Uganda, Sudan, Egypt, Kenya, Tanzania and Zambia.

Chamaeleo (Chamaeleo) **monachus** GRAY

1865 *Chamaeleon monachus* GRAY, Proc. zool Soc. London, **1864**: 470; plate 31. – Terra typica: Madagascar. – Holotype (by monotypy): BMNH 1946.8.21.98.
1911 *Chamaeleon monachus* – WERNER, Das Tierreich, **27**: 14.
1966 *Chamaeleo monachus* – MERTENS, Das Tierreich, **83**: 20.
1986 *Chamaeleo (Chamaeleo) monachus* – KLAVER & BÖHME, Bonn. zool. Monogr., **22**: 59.

Distribution: Socotra Island (= Suqutra), Yemen.

Chamaeleo (Chamaeleo) **namaquensis** SMITH*

?1789 *Chamaeleo capensis* BONNATERRE, Tabl. encyclop., Erpétol.: 32; plate 7, fig. 3. – Terra typica: not given. – Terra typica restricta: Cape Province, South Africa (fide MERTENS 1966: 21). – Type(s): not located.
1831 *Chaemelio namaquensis* SMITH, S. Afr. quart. J., **5**: 17. – Terra typica: Little Namaqualand, near the mouth of the Orange River, South Africa. – Syntype: MHNP 8017.
1845 *Chameleo tuberculiferus* GRAY, Cat. Lit. Brit. Mus.: 267. – Terra typica: Namaqualand, S.W. Africa. – Syntypes: BMNH 38.3.8.5–6 (2 specimens). – Syn.: fide DUMÉRIL & DUMÉRIL 1851: 32.
1911 *Chamaeleon namaquensis* – WERNER, Das Tierreich, **27**: 40; fig. 14.
1966 *Chamaeleo namaquensis* – MERTENS, Das Tierreich, **83**. 21.
1986 *Chamaeleo (Chamaeleo) namaquensis* – KLAVER & BÖHME, Bonn. zool. Monogr., **22**: 59.

Distribution: Southern Angola, Namibia and South Africa.

Chamaeleo (Chamaeleo) **quilensis** BOCAGE

1866 *Chamaeleo dilepis* var. *quilensis* BOCAGE, J. Sci., Lisboa, **1**: 42, 59. – Terra typica: Rio Quilo, Cabinda, Angola. – Holotype (by monotypy): MBL, lost by fire (fide SALDANHA 1978).
1887 *Chamaeleon parvilobus* BOULENGER, Cat. Liz. Brit. Mus., **3**: 449; plate 39, fig. 5. – Terra typica: Natal, Gaboon, Cameroon. – Syntypes: BMNH 53.10.29.23, 58.4.11.4, 69.3.22.1 and 76.3.8.3. – Syn.: fide WERNER 1902: 339.
1893 *Chamaeleon parrilobus* MOCQUARD (ex errore), C.R. Soc. philom., **19**: 5.
1895 *Chamaeleon, quilensis* – BOCAGE, Herpétol. Angola: 60; plate 8, fig. 3.
1897 *Chamaeleon quillensis* BOCAGE (ex errore), J. Acad. Sci., Lisboa, **4**: 198.
1911 *Chamaeleon dilepis quilensis* – WERNER, Das Tierreich, **27**: 13.
1966 *Chamaeleo quilensis* – MERTENS, Das Tierreich, **83**: 25.
1986 *Chamaeleo (Chamaeleo) quilensis* – KLAVER & BÖHME, Bonn. zool. Monogr., **22**: 59.

Distribution: Equatorial Guinea, Gabon, Congo, Zaire, Angola, Uganda, Kenya, Tanzania, Malawi, Mozambique, Zambia, Zimbabwe, S. Namibia and South Africa.

Chamaeleo (Chamaeleo) **senegalensis** DAUDIN

1802 *Chamaeleo senegalensis* DAUDIN, Hist, nat. Rept., **4**: 203. – Terra typica: Sénégal. – Holotype (by monotypy): MHNP 2224.
?1820 *Chamaeleon subcroceus* MERREM, Tent. Syst. Amph.: 162. – Terra typica: Africa. – Type(s): PUM, presumed lost (H.O. von Hagen, in litt. 8.IX.1994). – Syn.: fide MERTENS 1966: 27.
1825 *Chamaeleo gymnocephalus* CAUP (= KAUP; nomen sustitutum pro *Chamaeleo senegalensis* DAUDIN 1802), Isis (Oken), Leipzig **1825** (1): 592.
?1831 *Chameleon galeoratus* GRAY (nomen nudum) in: GRIFFITH, Anim. Kindgdom, **9** (Synops): 53.
?1843 *Chamaeleon leptopus* FITZINGER (nomen nudum), Syst. Rept.: 41.
1865 *Chamaeleon senegalensis* var.? *leiocephalus* GRAY, Proc. zool. Soc. London, **1864**: 471. – Terra typica: not given. – Type(s): BMNH; not registered as type(s). – Syn.: fide WERNER 1902: 12.
1887 *Chamaeleon liocephalus* BOETTGER (nomen substitutum pro *Chamaeleon senegalensis* var.? *leiocephalus* GRAY 1865), Ber. senckenb. naturf. Ges., **1887**: 58.
1893 *Chamaeleon licocephalus* STERNFELD (ex errore), Ergebn. zweit. dtsch. Zentral-Afr.-Exped. 1910–1911, **1**: 452.
1911 *Chamaeleon senegalensis* – WERNER, Das Tierreich, **27**: 12.
1966 *Chamaeleo senegalensis* – MERTENS, Das Tierreich, **87**: 27.
1966 *Chamaeleo senegalensis senegalensis* – MERTENS, Das Tierreich, **83**: 27.
1986 *Chamaeleo (Chamaeleo) senegalensis* – KLAVER & BÖHME, Bonn. zool. Monogr., **22**: 59.

Distribution: Arid W. Africa; from Senegal to N. Cameroon.

Chamaeleo (Chamaeleo) **zeylanicus** LAURENTI

1768 *Chamaeleo zeylanicus* LAURENTI, Synops. Rept.: 46. – Terra typica: not given. – Terra typica restricta: Ceylon (= Sri Lanka) (fide MERTENS 1969: 41). – Type(s): not located; based on Seba (1734) I; plate 82, fig. 3.
1823 *Chamaeleon zebra* BORY DE ST. VINCENT, Dict. Hist. nat., **3**: 97; Atlas: plate 121. – Terra typica: Indien. – Type(s): not located. – Syn. fide BOULENGER 1887: 445.
1843 *Chamaeleon coromandelicus* FITZINGER, Syst. Rept.: 41. – Terra typica: Indien. – Type(s): not located; not in NMW (H. GRILLITSCH, in litt. 7.II.1995). – Syn.: fide BOULENGER, 1887: 445.
1853 *Chamaeleo zeylonicus* JERDON (nomen substitutum pro *Chamaeleo zeylanicus* LAURENTI 1768), J. Asiat. Soc. Bengal, Calcutta, **22**: 466.
1865 *Chamaeleon vulgaris* var. *marmoratus* GRAY, Proc. zool. Soc. London, **1864**: 469. – Terra typica: Dukhun (= Deccan), India. – Syntypes: BMNH 1946.8.22.3–4 (2 specimens). – Syn.: fide BOULENGER 1887: 445.

1872 *Chamaeleo ceylonicus* STOLICZKA (nomen substitutum pro *Chamaeleo zeylanicus* LAURENTI 1768), Proc. asiat. Soc. Bengal, Calcutta, **1872**: 81.
1911 *Chamaeleon calcaratus* – WERNER, Das Tierreich, **27**: 11.
1959 *Chamaeleo chamaeleon zeylanicus* – HILLENIUS, Beaufortia, Amsterdam, **8**: 46.
1966 *Chamaeleo chamaeleon zeylanicus* – MERTENS, Das Tierreich, **83**: 11.
1986 *Chamaeleo (Chamaeleo) zeylanicus* – KLAVER & BÖHME, Bonn. zool. Monogr., **22**: 59.

Distribution: Sri Lanka, India and S. E. Pakistan.

Subgenus Trioceros SWAINSON*

1839 *Trioceros* SWAINSON, Nat. Hist. Fishes Amph. Rept., **2**: 347, 369. – Species typica (by monotypy): *Trioceros grayii* SWAINSON 1839 (= *Chameleo oweni* GRAY 1831).
1843 *Triceras* FITZINGER (non *Triceras* LOBARZEWSKI 1840 = Spongia), Syst. Rept.: 15, 42. – Species typica (by original designation): *Chameleon owenii* GRAY 1831.
1865 *Pterosaurus* GRAY (non *Pterosaurus* FITZINGER 1843 = Reptilia, Agamidae), Proc. zool. Soc. London, **1864**: 467, 473. – Species typica (by monotypy): *Chameleo cristatus* GRAY (= STUTCHBURY 1837).
1865 *Pterosaura* GRAY (ex errore), Proc. zool. Soc. London, **1864**: 467.
1865 *Ensirostris* GRAY, Proc. zool. Soc. London, **1864**: 467, 468, 478; plate 32, fig. 1. – Species typica (by monotypy): *Ensirostris melleri* GRAY 1865.
1994 *Triceros* NEČAS (ex errore), Herpetozoa, Vienna, **7**: 105; fig. 3.

Distribution: Tropical Africa, from Nigeria, Cameroon, Gabon, Zaire and Angola (Cabinda) to Sudan, Ethiopia, Somalia, Kenya, Tanzania, Malawi and Mozambique.

Chamaeleo (Trioceros) **affinis** RÜPPELL

1843 *Chamaeleon abyssinicus* FITZINGER (nomen nudum), Syst. Rept.: 42.
1845 *Chamaeleo affinis* RÜPPELL, Mus. senckenb., **3**: 300. – Terra typica: Abessinien (= Ethiopia). – Lectotypus: SMF 16402 (fide MERTENS 1967: 64).
1911 *Chamaeleon affinis* – WERNER, Das Tierreich, **27**: 27.
1966 *Chamaeleo affinis* – MERTENS, Das Tierreich, **83**: 6.
1986 *Chamaeleo (Trioceros) affinis* – KLAVER & BÖHME, Bonn. zool. Monogr., **22**: 59.

Distribution: Ethiopia.

Chamaeleo (Trioceros) **bitaeniatus** FISCHER*

1884 *Chamaeleo bitaeniatus* FISCHER, Jb. hamburg. Wiss. Anst., **1**: 23; plate 2, figs. 7A–B. – Terra typica: Naiwascha-See (= Lake Naivasha), Kenya. – Syntypes: ZMH-R02832–36 (5 specimens).
1887 *Chamaeleo bivittatus* MÜLLER (nomen nudum), Verh. naturf. Ges. Basel, **8**: 294.
1911 *Chamaeleon bitaeniatus* – WERNER, Das Tierreich, **27**: 14.
1911 *Chamaeleon bitaeniatus bitaeniatus* – WERNER, Das Tierreich, **27**: 15.
1916 *Chamaeleo bilineatus* SÉWERTZOFF (nomen substitutum pro *Chamaeleo bitaeniatus* FISCHER 1884), J. russ. de Zool., Petrograd, **1**: 669; plate 22–23.
1966 *Chamaeleo bitaeniatus* – MERTENS, Das Tierreich, **83**: 8.
1966 *Chamaeleo bitaeniatus bitaeniatus* – MERTENS, Das Tierreich, **83**: 8.
1986 *Chamaeleo (Trioceros) bitaeniatus* – KLAVER & BÖHME, Bonn. zool. Monogr., **22**: 59.

Distribution: Ethiopia, S. Sudan, Somalia, Kenya, Uganda, N. Tanzania, and N.E. Zaire.

Chamaeleo (Trioceros) **camerunensis** MÜLLER

1909 *Chamaeleon camerunensis* MÜLLER, Jb. nassau. Ver. Naturk., Wiesbaden, **62**: 111. – Terra typica: Dibongo near Edea, Cameroon. – Lectotype: ZSM 425/1909 (fide KLAVER & BÖHME 1992: 445).
1911 *Chamaeleon camerunensis* – WERNER, Das Tierreich, **27**: 36.
1965 *Chamaeleo montium camerunensis* – MERTENS, Bonn. zool. Beitr., **15** (1964): 219.
1966 *Chamaeleo montium camerunensis* – MERTENS, Das Tierreich, **83**: 20.
1986 *Chamaeleo (Trioceros) camerunensis* – KLAVER & BÖHME, Bonn. zool. Monogr., **22**: 59.

Distribution: Lowlands south of Mt. Cameroon and south of the Manengouba Mts., Cameroon.

Chamaeleo (Trioceros) **chapini** DE WITTE*

1964 *Chamaeleo chapini* DE WITTE, Amer. Mus. Novitates, **2192**: 1; fig. 1. – Terra typica: Lukolela, Terr. of Bikoro, Equator district of the province Equator, Congo (= Zaire). – Holotype (by orig. design.): AMNH 45885.
1966 *Chamaeleo chapini* – MERTENS, Das Tierreich, **83**: 11.
1986 *Chamaeleo (Trioceros) chapini* – KLAVER & BÖHME, Bonn. zool. Monogr., **22**: 59.

Distribution: Zaire and Gabon (fide KNOEPFFLER 1967: 245).

Chamaeleo (Trioceros) **cristatus** STUTCHBURY

1837 *Chamaeleon cristatus* STUTCHBURY, Trans. Linn. Soc., London, **17**: 361; plate 10. – Terra typica: Gaboon. – Holotype (by monotypy): CBMAG A715; presumed lost (fide KLAVER & BÖHME 1992: 446).
1911 *Chamaeleon cristatus* – WERNER, Das Tierreich, **27**: 33; fig. 10.
1966 *Chamaeleo cristatus* – MERTENS, Das Tierreich, **83**: 11.
1986 *Chamaeleo (Trioceros) cristatus* – KLAVER & BÖHME, Bonn. zool. Monogr., **22**: 59.

Distribution: Gabon, Nigeria, Cameroon, Central African Republic, Equatorial Guinea (Mbini and Bioko) and Congo (? Togo and ? Ghana, fide KLAVER & BÖHME 1992: 447).

Chamaeleo (Trioceros) **deremensis** MATSCHIE

1892 *Chamaeleon deremensis* MATSCHIE, S.Ber. Ges. naturf. Fr. Berlin, **1892**: 103. – Terra typica: Derema, Usambara Mts., Tanzania. – Syntypes: ZMB 11347 & 11352 (3 specimens).
1911 *Chamaeleon deremensis* – WERNER, Das Tierreich, **27**: 34.
1966 *Chamaeleo deremensis* – MERTENS, Das Tierreich, **83**: 12.
1986 *Chamaeleo (Trioceros) deremensis* – KLAVER & BÖHME, Bonn. zool. Monogr., **22**: 59.

Distribution: Usambara Mts. and Uluguru Mts. (fide BARBOUR & LOVERIDGE 1928: 176), Tanzania.

Chamaeleo (Trioceros) **eisentrauti** MERTENS

1968 *Chamaeleo eisentrauti* MERTENS, Bonn. zool. Beitr., **19**: 70; figs. 1–2, plate. – Terra typica: Dikume, Rumpi Mts. (1150 m alt.), Cameroon. – Holotype (by orig. design.): ZFMK 5781.
1986 *Chamaeleo (Trioceros) eisentrauti* – KLAVER & BÖHME, Bonn. zool. Monogr., **22**: 59.

Distribution: Known only from the type locality.

Chamaeleo (Trioceros) **ellioti** GÜNTHER

1895 *Chamaeleon ellioti* GÜNTHER, Ann. Mag. nat. Hist., London, **15**: 524; plate 21, fig. A. – Terra typica: Kavirondo (3900–4000 ft. alt.), and foot of Mount Ruwenzori (5000–6000 ft. alt.). – Terra typica restricta (fide LOVERIDGE, 1944: 147): Bugoye, east foot of Ruwenzori Mts., Uganda. – Syntypes: BMNH 1946.8.20.99, 1946.8.21.1–4 & 1946.8.22.17–20 (9 specimens).
1897 *Chamaeleon bitaeniatus* var. *ellioti* – TORNIER, Kriecht. Dtsch.-Ost.-Afr.: 54.
1911 *Chamaeleon bitaeniatus ellioti* – WERNER, Das Tierreich, **27**: 15; fig. 3.
1922 *Chamaeleon bequaerti* DE WITTE, Rev. zool. afr., Bruxelles, **10**: 69; plate 2, fig. 1. – Terra typica: Beni, Kivu district, Zaire. – Holotype (by monotypy): MRAC 1971. – Syn.: fide DE WITTE 1933: 120.
1966 *Chamaeleo bitaeniatus ellioti* – MERTENS, Das Tierreich, **83**: 8.
1980 *Chamaeleo ellioti* – BÖHME & KLAVER, Amphibia-Reptilia, **1**: 15.
1986 *Chamaeleo (Trioceros) ellioti* – KLAVER & BÖHME, Bonn. zool. Monogr., **22**: 59.

Distribution: S. Sudan, W. Kenya, Uganda, Rwanda, Burundi, E. Zaire and W. Tanzania.

Chamaeleo (Trioceros) **feae** BOULENGER

1906 *Chamaeleon feae* BOULENGER, Ann. Mus. civ. Stor. nat., Genova, **2**: 207; 2 figs. – Terra typica: Moka (1300–1500 m alt.), Fernando Poo (= Bioko, Equatorial Guinea). – Lectotype: MCSN-EE 28165 (fide KLAVER & BÖHME 1992: 452).
1911 *Chamaeleon feae* – WERNER, Das Tierreich, **27**: 37.
1965 *Chamaeleo montium feae* – MERTENS, Bonn. zool. Beitr., **15** (1964): 212, 214.
1966 *Chamaeleo montium feae* – MERTENS, Das Tierreich, **83**: 21.
1986 *Chamaeleo (Trioceros) feae* – KLAVER & BÖHME, Bonn. zool. Monogr., **22**: 59.

Distribution: Fernando Po (= Bioko), Equatorial Guinea.

Chamaeleo (Trioceros) **fuelleborni** TORNIER

1900 *Chamaeleon fülleborni* TORNIER, Zool. Jb., Syst., **13**: 614; fig. H. – Terra typica: Nonde – Nike (= Konde and Nyika), Kungura-Berg (2500 m alt.) and Ngosi- oder Peroto-Berge (2200 m alt.), Tanzania. – Syntypes: ZMB 15708–15710 (3 specimens) and NMW 23384 (1 specimen).
1911 *Chamaeleon fuelleborni* – WERNER, Das Tierreich, **27**: 25.
1966 *Chamaeleo fuelleborni* – MERTENS, Das Tierreich, **83**: 15.
1986 *Chamaeleo (Trioceros) fuelleborni* – KLAVER & BÖHME, Bonn. zool. Monogr., **22**: 59.

Distribution: Ngosi vulcano, Poroto Mts., Ukonde, S. Tanzania.

Chamaeleo (Trioceros) **goetzei** TORNIER

1899 *Chamaeleon goetzei* TORNIER, Zool. Anz., **22**: 413; plate 2, fig. 3. – Terra typica: Uzungwe Mts., Uhehe, Tanzania. – Holotype (by monotypy): ZMB 14972.
1966 *Chamaeleo goetzei* – MERTENS, Das Tierreich, **83**: 16.
1986 *Chamaeleo (Trioceros) goetzei* – KLAVER & BÖHME, Bonn. zool. Monogr., **22**: 59.

Distribution: Malawi and Tanzania.

Chamaeleo (Trioceros) goetzei goetzei TORNIER

1903 *Chamaeleon goetzii* – MONK (ex errore), The Zoologist, **7**: 326.
1911 *Chamaeleon goetzei* – WERNER, Das Tierreich, **27**: 17.

1953 *Chamaeleo goetzei goetzei* – LOVERIDGE, Bull. Mus. comp. Zool., **110**: 190.
1966 *Chamaeleo goetzei goetzei* – MERTENS, Das Tierreich, **83**: 16.

Distribution: Uzungwe Mts., Ubena Mts., Ukinga Mts., Rungwe Mts. and Poroto Mts., Tanzania.

Chamaeleo (Trioceros) goetzei nyikae LOVERIDGE

1953 *Chamaeleo goetzei nyikae* LOVERIDGE, Bull. Mus. comp. Zool., **110**: 189. – Terra typica: Nyika Plateau, directly above Nchenachena (approx, 7500 ft. alt.), Nyassaland (= Malawi). – Holotype (by orig. design.): MCZ 50554.
1966 *Chamaeleo goetzei nyikae* – MERTENS, Das Tierreich, **83**: 16.

Distribution: Nyika Plateau, Malawi.

Chamaeleo (Trioceros) **harennae** LARGEN

1995 *Chamaeleo harennae* LARGEN, Trop. Zool., **4**: 334; figs. 1–2. – Terra typica: Katcha, Harenna Forest, Bale Mountains, Ethiopia. – Holotype (by orig. design.): LIV 1986.212.246.

Distribution: Bale Mountains, Ethiopia.

Chamaeleo (Trioceros) **hoehnelii** STEINDACHNER

1891 *Chamael(e)on höhnelii* STEINDACHNER, Anz. Akad. Wiss. Wien, **28**: 141 / S.Ber. Akad. Wiss. Wien, math.-naturw. Kl., Abt. 1, **100**: 309; plate 1, figs. 1–1A. – Terra typica: Leikipia (6000 ft. alt.), Kenya. – Holotype (by monotypy): NMW 23380.
1891 *Chamaeleon leikipiensis* STEINDACHNER, Anz. Akad. Wiss. Wien, **28**: 142 / S.Ber. Akad. Wiss. Wien, math.-naturw. Kl., Abt. 1, **100**: 311; plate 1, figs. 2–2A. – Terra typica: Leikipia (6000 ft. alt.), Kenya. – Holotype (by monotypy): NMW 7716. – Syn.: fide BOULENGER 1892: 72.
1911 *Chamaeleon bitaeniatus höhneli* – WERNER, Das Tierreich, **27**: 15.
1911 *Chamaeleon bitaeniatus leikipiensis* – WERNER, Das Tierreich, **27**: 15.
1912 *Chamaeleon bitaeniatus bergeri* STERNFELD, S.Ber. Ges. naturf. Fr. Berlin, **1912**: 380; plate 14, fig. 16; plate 16, fig. 31/Wiss. Ergebn. dtsch. Zentral-Afr.-Exped. 1907–1908, **4** (zool., **2**): 252; plate 9, fig. 5. – Terra typica: Sirgoi, südlich von Ravine, later corrected in: Sergoit, N. of Eldama Ravine Station (fide RAND 1963: 9), Kenya. – Type(s): not located. – Syn.: fide LOVERIDGE 1929: 89.
1913 *Chamaeleon bitaeniatus hoeneli* GERMERSHAUSEN (ex errore), S.Ber. Ges. naturf. Fr. Berlin, **1913**: 469, 517.
1935 *Chamaeleon bitaeniatus altaeelgonis* LOVERIDGE, Bull. Mus. comp. Zool., **79**: 15. – Terra typica: Kaburomi (10500 ft. alt.), Mt. Elgon, Uganda. – Holotype (by orig. design.): MCZ 40274. – Syn.: fide RAND 1963: 10.
1966 *Chamaeleo hoehnelii* – MERTENS, Das Tierreich, **83**: 17.
1966 *Chamaeleo hoehnelii hoehnelii* – MERTENS, Das Tierreich, **83**: 17.
1966 *Chamaeleo hoehnelii altaeelgonis* – MERTENS, Das Tierreich, **83**: 17.
1986 *Chamaeleo (Trioceros) hoehnelii* – KLAVER & BÖHME, Bonn. zool. Monogr., **22**: 59.

Distribution: Highlands of E. Uganda and Kenya.

Chamaeleo (Trioceros) **incornutus** LOVERIDGE

1932 *Chamaeleon incornutus* LOVERIDGE, Bull. Mus. comp. Zool., **72**: 380. – Terra typica: Madehani (7000 ft. alt.), Ukinga Mts., Tanzania. – Holotype (by orig. design.): MCZ 31350.

1966 *Chamaeleo incornutus* – MERTENS, Das Tierreich, **83**: 17.
1986 *Chamaeleo (Trioceros) incornutus* – KLAVER & BÖHME, Bonn. zool. Monogr., **22**: 59.

Distribution: Ukinga Mts., Ubena Mts., Rungwe Mts. and Poroto Mts., Tanzania.

Chamaeleo (Trioceros) **ituriensis** SCHMIDT*

1912 *Chamaeleon johnstoni affinis* STERNFELD (non *Chamaeleo affinis* RÜPPELL 1843), Wiss. Ergebn. dtsch. Zentr.-Afr.-Exped., 1907–1908, **4**: (Zool. 2): 262; plate 7, fig. 3. – Terra typica: Randberge am N.-W. Ufer des Tanganyika und Irumu-Mavambi-Urwald, Ost-Afrika. – Syntypes: ZMB 22377 (only one of the two original specimens). – Syn.: fide SCHMIDT 1919: 591.
1919 *Chamaeleon ituriensis* SCHMIDT, Bull. Amer. Mus. nat. Hist. **39**: 589; plate 31, fig. 2; plate 32, figs. 1–5. – Terra typica: Medje, Zaire. – Holotype (by orig. design.): AMNH 11490.
1942 *Chamaeleo johnstoni ituriensis* – LOVERIDGE, Bull. Mus. comp. Zool., **91**: text to plate 6, fig. 1.
1966 *Chamaeleo johnstoni ituriensis* – MERTENS, Das Tierreich, **83**: 19.

Distribution: E. Zaire.

Chamaeleo (Trioceros) **jacksonii** BOULENGER

1896 *Chamaeleon jacksonii* BOULENGER, Ann. Mag. nat. Hist., London, **17**: 376. – Terra typica: Uganda, later amended to Kikuyu, near Nairobi, Kenya (fide MERTENS 1966: 18). – Holotype (by monotypy): BMNH 1946.8.21.81.
1966 *Chamaeleo jacksonii* – MERTENS, Das Tierreich, **83**: 18.
1986 *Chamaeleo (Trioceros) jacksonii* – KLAVER & BÖHME, Bonn. zool. Monogr., **22**: 59.

Distribution: Kenya and Tanzania.

Chamaeleo (Trioceros) jacksonii jacksonii BOULENGER*

1904 *Chamaeleon jacksoni* var. *vauerescecae* TORNIER, Zool. Jb., Syst., **19**: 176. – Terra typica: Nairobi in Kikuju (7000 ft. alt.), Kenya. – Syntypes: ZMB 18110–18114 (17 specimens and numerous juveniles). – Syn.: fide MERTENS 1949: 199.
1911 *Chamaeleon jacksoni* – WERNER, Das Tierreich, **27**: 26.
1911 *Chamaeleon jacksoni jacksoni* – WERNER, Das Tierreich, **27**: 26.
1911 *Chamaeleon jacksoni vauerescecae* – WERNER, Das Tierreich, **27**: 26.
1966 *Chamaeleo jacksonii jacksonii* – MERTENS, Das Tierreich, **83**: 18.

Distribution: Highlands of Kenya.

Chamaeleo (Trioceros) jacksonii merumontanus RAND

1958 *Chamaeleo jacksoni merumontana* RAND, Breviora, **99**: 1. – Terra typica: Laikinoi, near Arusha, Mt. Meru, Arusha district (7500 ft. alt.), Tanzania. – Holotype (by orig. design.): MCZ 56152.
1966 *Chamaeleo jacksonii merumontanus* – MERTENS, Das Tierreich, **83**: 18.

Distribution: Mt. Meru (7500–9000 ft. alt.), Tanzania.

Chamaeleo (Trioceros) jacksonii xantholophus EASON, FERGUSON & HEBRARD*

1988 *Chamaeleo jacksonii xantholophus* EASON, FERGUSON & HEBRARD, Copeia, **1988**: 586; fig. 1. – Terra typica: 3 km south of Meru, Meru district, Kenya. – Holotype (by orig. design.): NMK L/1796.

Distribution: Southern, eastern and north-eastern face of Mt. Kenya, Kenya and introduced in Hawaii (fide MCKEOWN 1991: 49).

Chamaeleo (Trioceros) **johnstoni** BOULENGER*

1901 *Chamaeleon johnstoni* BOULENGER, Proc. zool. Soc. London, **1901**: 136; plate 13. – Terra typica: Ruwenzori (6000 ft. alt.), Uganda. – Syntypes: BMNH 1946.8.21.36–40 (5 specimens).
1911 *Chamaeleon graueri* STEINDACHNER, Anz. Akad. Wiss. Wien, **48**: 177. – Terra typica: Nordwestliches Ufer des Tanganyika-Sees and Bukoba, Ost-Afrika. – Syntypes: NMW 8019:1 (several specimens). – Syn.: fide STERNFELD 1912: 260.
1911 *Chamaeleon johnstoni* – WERNER, Das Tierreich, **27**: 35.
1911 *Chamaeleon johnstoni johnstoni* – WERNER, Das Tierreich, **27**: 35.
1911 *Chamaeleon johnstoni graueri* – WERNER, Das Tierreich, **27**: 35.
1926 *Chamaeleon laevigularis* MÜLLER, Zool. Anz., **65**: 197. – Terra typica: East London, South Africa. – Holotype (by monotypy): ZSM 139/1925. – Syn.: fide HILLENIUS 1963: 44.
1942 *Chamaeleo johnstoni johnstoni* – LOVERIDGE, Bull. Mus. comp. Zool. **91**: plate 6 (text), fig. 1.
1951 *Chamaeleo johnstoni crenulatus* LAURENT, Rev. zool. bot. Afr., Bruxelles, **44**: 360. – Terra typica: May ya Moto (2450 m alt.), Terr. de Mwenge, Kivu, Zaire. – Holotype (by orig. design.): MRAC 26.365. – Syn.: fide De Witte 1965: 158, 162.
1966 *Chamaeleo johnstoni* – MERTENS, Das Tierreich, **83**: 18.
1966 *Chamaeleo johnstoni johnstoni* – MERTENS, Das Tierreich, **83**: 18.
1986 *Chamaeleo (Trioceros) johnstoni* – KLAVER & BÖHME, Bonn. zool. Monogr., **22**: 59.

Distribution: Uganda, Rwanda, Burundi and E. Zaire.

Chamaeleo (Trioceros) **kinetensis** SCHMIDT

1943 *Chameleo bitaeniatus kinetensis* SCHMIDT, Field Mus. Nat. Hist., Zool. Ser., **24**: 336. – Terra typica: Mount Kineti, Imatong Mts. (10458 ft. alt.), Sudan. – Holotype (by orig. design.): FMNH 34484.
1963 *Chamaeleo kinetensis* – RAND, Bull. Mus. comp. Zool., **130**: 21.
1966 *Chamaeleo bitaeniatus kinetensis* – MERTENS, Das Tierreich, **83**: 8.
1980 *Chamaeleo kinetensis* – BÖHME & KLAVER, Amphibia-Reptilia, **1**: 4; figs. 1A–B, 2, 6.
1986 *Chamaeleo (Trioceros) kinetensis* – KLAVER & BÖHME, Bonn. zool. Monogr., **22**: 59.

Distribution: Mt. Kineti, Imatong Mts., Sudan.

Chamaeleo (Trioceros) **laterispinis** LOVERIDGE

1932 *Chamaeleon laterispinis* LOVERIDGE, Bull. Mus. comp. Zool., **72**: 381. – Terra typica: Kigogo, near Mufindi, Uzungwe Mts., Tanzania. – Holotype (by orig. design.): MCZ 31386.
1966 *Chamaeleo laterispinis* – MERTENS, Das Tierreich, **83**: 19.
1982 *Chamaeleo laterispinis laterispinis* – BÖHME, Bonn. Zool. Beitr., **33**: 350; figs. 1–3.
1982 *Chamaeleo laterispinis brookesiaeformis* BÖHME, Bonn. Zool. Beitr., **33**: 350; figs. 1–3. – Terra typica: Neu Mufindi (= Kibau Iyayi), S.W. Uzungwe Mts., Tanzania. – Holotype (by orig. design.): ZFMK 30785. – Syn.: fide BÖHME 1987: 30.
1986 *Chamaeleo (Trioceros) laterispinis* – KLAVER & BÖHME, Bonn. zool. Monogr., **22**: 59.

Distribution: Uzungwe Mts., Tanzania.

Chamaeleo (Trioceros) **marsabitensis** TILBURY

1991 *Chamaeleo (Trioceros) marsabitensis* TILBURY, Trop. Zool., **4**: 159; figs. 1–2. – Terra typica: Marsabit Mountain forest, Kenya. – Holotype (by orig. design.): BMNH 1988.406.

Distribution: Known only from the type locality.

Chamaeleo (Trioceros) **melleri** (GRAY)

1865 *Ensirostris melleri* GRAY, Proc. zool. Soc. London, **1864**: 478; plate 32, fig. 1. – Terra typica: East Africa, on the mountains of the interior. – Terra typica restricta (fide LOVERIDGE, 1953: 187): Zomba, Nyassaland (= Malawi). – Holotype (by monotypy): BMNH 64.10.29.1.
1899 *Chamaeleon welleri* (ex errore) – TORNIER, Zool. Anz., **22**: 258.
1911 *Chamaeleon melleri* – WERNER, Das Tierreich, **27**: 37.
1966 *Chamaeleo melleri* – MERTENS, Das Tierreich, **83**: 19.
1986 *Chamaeleo (Trioceros) melleri* KLAVER & BÖHME, Bonn. zool. Monogr., **22**: 59.

Distribution: Malawi, Tanzania and N. Mozambique.

Chamaeleo (Trioceros) **montium** BUCHHOLZ

1874 *Chamaeleo montium* BUCHHOLZ, M.Ber. Akad. Wiss. Berlin, **1874**: 88. – Terra typica: Bonjongo, Mt. Cameroon, Cameroon. – Lectotype: ZMB 8025.1 (fide KLAVER & BÖHME 1992: 455).
1911 *Chamaeleon montium* – WERNER, Das Tierreich, **27**: 35; fig. 12.
1938 *Chamaeleo montium montium* – MERTENS, Abh. senckenb. naturf. Ges., **442**: 41.
1938 *Chamaeleo montium grafi* MERTENS, Abh. senckenb. naturf. Ges., **442**: 41; plate 9, figs. 40–41. – Terra typica: Mongonge (550 m alt.), Mt Cameroon. – Holotype (by orig. design.): SMF 22460. – Syn.: fide KLAVER & BÖHME 1992: 458.
1966 *Chamaeleo montium* – MERTENS, Das Tierreich, **82**: 20.
1966 *Chamaeleo montium montium* – MERTENS, Das Tierreich, **83**: 20.
1966 *Chamaeleo montium grafi* – MERTENS, Das Tierreich, **83**: 21.
1986 *Chamaeleo (Trioceros) montium* – KLAVER & BÖHME, Bonn. zool. Monogr., **22**: 59.
1987 *Chamaeleo monticum* RIEPPEL (ex errore), Zool. J. Linn. Soc., London, **89**: 54.

Distribution: Mt. Cameroon, Rumpi Hills, Mt. Kupe, Manengouba Mts. and Mamfe district, Cameroon.

Chamaeleo (Trioceros) **oweni** GRAY

1831 *Chameleo oweni* GRAY, Zool. Misc., London, **1**: 7; plate 4. – Terra typica: Fernando Poo (= Bioko), Equatorial Guinea. – Syntypes: BMNH 1946.8.21.48–49 (2 specimens).
1831 *Chameleon owenii* – GRAY (ex errore) in: GRIFFITH, Anim. Kingdom, 9 (Synops.): 54.
1836 *Chamaeleo tricornis* DUMÉRIL & BIBRON, Erpétol gén., **3**: 227. – Terra typica: Fernando Poo. – Type(s): not located. – Syn.: fide GRAY 1845: 269.
1838 *Chamaeleon bibroni* MARTIN, Proc. zool. Soc. London, **1838**: 65. – Terra typica: Fernando Poo. – Holotype (by monotypy): BMNH 1946.8.21.66. – Syn.: fide GRAY 1845: 269.
1839 *Trioceros grayii* SWAINSON, (nomen substitutum pro *Chameleo oweni* GRAY 1831), Nat. Hist. Fishes Amph. Rept., **2**: 369.
1906 *Chamaeleon unicornis* MOCQUARD, Bull. Mus. Hist. nat., Paris, **12**: 464. – Terra typica: Ngomo, dans le bas Ogoué, Gaboon. – Holotype (by monotypy): MHNP 06.173. – Syn.: fide DE WITTE 1965: 184.
1911 *Chamaeleon owenii* – WERNER, Das Tierreich, **27**: 34; fig. 11.
1911 *Chamaeleon unicornis* – WERNER, Das Tierreich, **27**: 34.
1912 *Chamaeleon oweni* subsp. *cristata* STERNFELD (non *Chamaeleo cristatus* STUTCHBURY 1837), Wiss. Ergebn. dtsch. Zentral-Afr.-Exped. 1907–1908: 259. – Terra typica: Landana in Loango, nördlich der Kongomündung, Zaire. – Holotype (by monotypy): ZMB 23924. – Syn.: fide SCHMIDT 1919: 587.
1913 *Chamaeleo michelli* MÜLLER, Zool. Anz., **41**: 230. – Terra typica: Station Gô am Rulei, unterer Uelle District, Zaire. – Holotype (by monotypy): ZSM 286/1912. – Syn.: fide SCHMIDT 1919: 587.
1919 *Chamaeleon mitchelli* SCHMIDT (ex errore), Bull. Amer. Mus. Nat. Hist., **39**: 584.
1959 *Chamaeleo oweni oweni* – HILLENIUS, Beaufortia, Amsterdam, **8**: 54.

1959 *Chamaeleo oweni unicornis* – HILLENIUS, Beaufortia, Amsterdam, **8**: 54.
1966 *Chamaeleo oweni* – MERTENS, Das Tierreich, **83**: 21.
1986 *Chamaeleo (Trioceros) oweni* – KLAVER & BÖHME, Bonn. zool. Monogr., **22**: 59.

Distribution: Nigeria, Cameroon, Equatorial Guinea (Mbini and Bioko), Gabon, Congo, Zaire and Angola (Cabinda).

Chamaeleo (Trioceros) **pfefferi** TORNIER

1900 *Chamaeleon pfefferi* TORNIER, Zool. Anz., **23**: 21; fig. – Terra typica: Nyassosso, Nkossogebirge (= Nyasoso, Mt. Kupe), Cameroon. – Holotype (by monotypy): ZMB 15050.
1911 *Chamaeleon pfefferi* – WERNER, Das Tierreich, **27**: 23.
1966 *Chamaeleo pfefferi* – MERTENS, Das Tierreich, **83**: 23.
1986 *Chamaeleo (Trioceros) pfefferi* – KLAVER & BÖHME, Bonn. zool. Monogr., **22**: 59.

Distribution: Mt. Kupe and Manengouba Mts., Cameroon (fide WILD 1993).

Chamaeleo (Trioceros) **quadricornis** TORNIER

1899 *Chamaeleo quadricornis* TORNIER, Zool. Anz., **22**: 259; fig. 2. – Terra typica: Cameroon. – Terra typica restricta (fide KLAVER 1981: 44): Manengouba Mts., Cameroon. – Lectotype: ZMB 14957 (fide MERTENS 1959: 114, non BÖHME & KLAVER 1981: 317; fide KLAVER & BÖHME 1992: 463).
1986 *Chamaeleo (Trioceros) quadricornis* – KLAVER & BÖHME, Bonn. zool. Monogr., **22**: 59.

Distribution: Highlands of Cameroon and Nigeria.

Chamaeleo (Trioceros) quadricornis quadricornis TORNIER

1911 *Chamaeleon quadricornis* – WERNER, Das Tierreich, **27**: 36.
1966 *Chamaeleo quadricornis* – MERTENS, Das Tierreich, **83**: 25.
1981 *Chamaeleo quadricornis quadricornis* – BÖHME & KLAVER, Amphibia-Reptilia **1**: 315.

Distribution: Manengouba Mts. and Mt. Kupe, Cameroon.

Chamaeleo (Trioceros) quadricornis gracilior BÖHME & KLAVER

1981 *Chamaelo quadricornis gracilior* BÖHME & KLAVER, Amphibia-Reptilia, **1**: 317; figs. 4, 5, 8A. – Terra typica: Mt. Lefo (1800 m alt.), Cameroon. – Holotype (by orig. design.): ZFMK 15291.

Distribution: Bamboutos Mts., Mt. Lefo, Mt. Oku, Cameroon and Obudu plateau, Nigeria.

Chamaeleo (Trioceros) **rudis** BOULENGER*

1906 *Chamaeleon rudis* BOULENGER, Ann. Mag. nat. Hist., London, **18**: 473. – Terra typica: Mt. Ruwenzori, Uganda. – Holotype (by monotypy): BMNH 1946.8.21.6.
1911 *Chamaeleon bitaeniatus rudis* – WERNER, Das Tierreich, **27**: 15.
1912 *Chamaeleon bitaeniatus graueri* STERNFELD (non *Chamaeleon graueri* STEINDACHNER 1911), S.Ber. Ges. naturf. Fr. Berlin, **1912**: 380; plate 15, figs. 19–21; plate 17, fig. 33 / Wiss. Ergebn. dtsch. Zentral-Afr.-Exped. 1907–1908, **4**: 250; plate 7, fig. 1. – Terra typica: Rugege- und Bugoie-Wälder, Ninagongo (2500–3000 m alt.), Ruwenzori Mts., Rwanda. – Syntypes: ZMB 22336 (4 specimens) and NMW 23379: 1,2 (2 specimens). – Syn.: fide RAND 1963: 17.
1912 *Chamaeleon bitaeniatus tornieri* STERNFELD (non *Chamaeleon tornieri* WERNER 1902), S.Ber. Ges. naturf. Fr. Berlin, **1912**: 383; plate 17, fig. 35. – Terra typica: Lendu Plateau, Ituri district, Zaire. – Holotype (by monotypy): ZMB 17392. – Syn.: fide RAND 1963: 17.

1933 *Chamaeleon burgeoni* De Witte, Rev. zool. bot. Afr., Bruxelles, **24**: 120; fig. – Terra typica: Mombasa near Lubero, Kivu district, Zaire. – Holotype (by orig. design.): MRAC 8979. – Syn.: fide Rand 1963: 18.
1963 *Chamaeleo rudis rudis* – Rand, Bull. Mus. comp. Zool., **130**: 15.
1963 *Chamaeleo rudis sternfeldi* Rand, Bull. Mus. comp. Zool., **130**: 18. – Terra typica: Laikinae, Mt. Meru (7500 ft. alt.), Arusha district, Tanzania. – Holotype (by orig. design.): MCZ 56173. – Syn.: fide Broadley & Howell 1991: 13.
1966 *Chamaeleo rudis* – Mertens, Das Tierreich, **83**: 26.
1966 *Chamaeleo rudis rudis* – Mertens, Das Tierreich, **83**: 26.
1966 *Chamaeleo rudis sternfeldi* – Mertens, Das Tierreich, **83**: 26.
1986 *Chamaeleo (Trioceros) rudis* – Klaver & Böhme, Bonn. zool. Monogr., **22**: 59.

Distribution: Burundi, Ethiopia, Rwanda, Kenya, Tanzania, W. Uganda and E. Zaire.

Chamaeleo (Trioceros) **schoutedeni** Laurent

1952 *Chamaeleo bitaeniatus schoutedeni* Laurent, Rev. zool. bot. Afr., Bruxelles, **46**: 20. – Terra typica: Vallée de la Kabumbe, région du Mont Kabobo (2400 m alt.), Terr. d'Albertville, Zaire. – Holotype (by orig. design.): MRAC 2186.
1963 *Chamaeleo rudis schoutedeni* – Rand, Bull. Mus. comp. Zool., **130**: 18.
1966 *Chamaeleo rudis schoutedeni* – Mertens, Das Tierreich, **83**: 26.
1980 *Chamaeleo schoutedeni* – Böhme & Klaver, Amphibia-Reptilia, **1**: 16.
1986 *Chamaeleo (Trioceros) schoutedeni* – Klaver & Böhme, Bonn. zool. Monogr., **22**: 59.

Distribution: E. Zaire and Rwanda.

Chamaeleo (Trioceros) **schubotzi** Sternfeld

1912 *Chamaeleon bitaeniatus schubotzi* Sternfeld, S.Ber. Ges. naturf. Fr. Berlin, **1912**: 381; plate 15, fig. 24; plate 17, fig. 34. / Wiss. Ergebn. dtsch. Zentral-Afr.-Exped., **4**: 252; plate 7, fig. 2; plate 9, fig. 6. – Terra typica: Mt. Kenya (4000 m alt.), Kenya. – Holotype (by monotypy): ZMB 15409.
1963 *Chamaeleo schubotzi* – Rand, Bull. Mus. comp. Zool., **130**: 21.
1966 *Chamaeleo rudis schubotzi* – Mertens, Das Tierreich, **83**: 26.
1980 *Chamaeleo schubotzi* – Böhme & Klaver, Amphibia-Reptilia, **1**: 16.
1986 *Chamaeleo (Trioceros) schubotzi* – Klaver & Böhme, Bonn. zool. Monogr., **22**: 59.

Distribution: Known only from the type locality.

Chamaeleo (Trioceros) **tempeli** Tornier*

1899 *Chamaeleon tempeli* Tornier, Zool. Anz., **22**: 411; plate 2, figs. 2, 4. – Terra typica: Uzungwe Mts., Uhehe, Tanzania. – Syntypes: ZMB 14965 (2 specimens).
1900 *Chamaeleon tempeli* var. *wolffi* Tornier, Zool. Jb., Syst., **13**: 614; fig. G. – Terra typica: Tardalla (= Tandala), Ukinga Mts., Tanzania. – Holotype (by monotypy): ZMB 15782. – Syn.: fide Loveridge 1933a: 199.
1900 *Chamaeleon tempeli* var. *wolfi* Tornier (ex errore), Zool. Jb., Syst., **13**: 614.
1911 *Chamaeleon tempeli* – Werner, Das Tierreich, **27**: 25.
1911 *Chamaeleon tempeli tempeli* – Werner, Das Tierreich, **27**: 25.
1911 *Chamaeleon tempeli wolffi* – Werner, Das Tierreich, **27**: 25.
1933 *Chamaeleon tempeli wolfi* Loveridge (ex errore), Bull. Mus. comp. Zool., **74**: 336.
1966 *Chamaeleo tempeli* – Mertens, Das Tierreich, **83**: 28.
1986 *Chamaeleo (Trioceros) tempeli* – Klaver & Böhme, Bonn. zool. Monogr., **22**; 59.

Distribution: Uzungwe Mts., Ubena Mts., Ukinga Mts. and Ufipa (= Fip between Lake Rukwa and Lake Tanganyika; fide Tornier 1902: 588), Tanzania.

Chamaeleo (Trioceros) **tremperi** NEČAS

1994 *Chamaeleo (Trioceros) tremperi* NEČAS, Herpetozoa, Vienna, **7**: 104; fig. 3, table 1. – Terra typica: Eldana River Station, Kenya. – Holotype (by orig. design.): NMW 7880.1.

Distribution: Known only from the type locality.

Chamaeleo (Trioceros) **werneri** TORNIER

1899 *Chamaeleon werneri* TORNIER, Zool. Anz., **22**: 258; fig. 1. – Terra typica: Maschona-Gebiet, Deutsch-Ost-Afrika, later corrected in: Uhehe, Uzungwe Mts., Tanzania (fide TORNIER 1899: 408). – Holotype (by monotypy): ZMB 14944.
1911 *Chamaeleon werneri* – WERNER, Das Tierreich, **27**: 38.
1932 *Chamaeleon werneri werneri* – LOVERIDGE, Bull. Mus. comp. Zool., **72**: 379.
1932 *Chamaeleon werneri dabagae* LOVERIDGE, Bull. Mus. comp. Zool., **72**: 379. – Terra typica: near Dabaga, Uzungwe Mts., Tanzania. – Holotype (by orig. design.): MCZ 31344. – Syn.: fide LOVERIDGE, 1957b: 4, 1957a: 205.
1966 *Chamaeleo werneri* – MERTENS, Das Tierreich, **83**: 28.
1986 *Chamaeleo (Trioceros) werneri* – KLAVER & BÖHME, Bonn. zool. Monogr., **22**: 59.

Distribution: Uzungwe Mts., Uluguru Mts. (fide LOVERIDGE 1957b: 4) and Ufipa (= Fip, between Lake Rukwa and Lake Tanganyika; fide TORNIER 1902: 588), Tanzania.

Chamaeleo (Trioceros) **wiedersheimi** NIEDEN

1910 *Chamaeleon wiedersheimi* NIEDEN, Arch. Naturgesch., **76**: 239. – Terra typica: Gendero-Gebirge (= Tchabal Mbabo) and Tsch'a (Bekom), Bamenda district, Cameroon. – Terra typica restricta (fide KLAVER & BÖHME 1992: 467): Tchabal Mbabo, Cameroon. – Lectotype: ZMB 21873 (fide KLAVER & BÖHME 1992: 467).
1986 *Chamaeleo (Trioceros) wiedersheimi* – KLAVER & BÖHME, Bonn. zool. Monogr., **22**: 59.

Distribution: Highlands of Cameroon and Nigeria.

Chamaeleo (Trioceros) wiedersheimi wiedersheimi NIEDEN

1911 *Chamaeleon wiedersheimi* – WERNER, Das Tierreich, **27**: 16.
1922 *Chamaeleon serratus* MERTENS, Zool. Anz., **54**: 191; fig. – Terra typica: Südkamerun. – Holotype (by orig. design.): SMNS 4640; presumed lost (fide KLAVER & BÖHME 1992: 267). – Syn.: fide MERTENS 1940: 240.
1966 *Chamaeleo wiedersheimi* – MERTENS, Das Tierreich, **83**: 29.
1992 *Chamaeleo wiedersheimi wiedersheimi* – KLAVER & BÖHME, Bonn. zool. Beitr., **43**: 467.

Distribution: Bamenda-Banso Highlands, Cameroon and Obudu Plateau, Mambila Mts. and Gotel Mts., Nigeria.

Chamaeleo (Trioceros) wiedersheimi perreti KLAVER & BÖHME

1992 *Chamaeleo (Trioceros) wiedersheimi perreti* KLAVER & BÖHME, Bonn. zool. Beitr., **43**: 469; fig. 5, plate 1. – Terra typica: Nsoung, Manengouba Mts., Cameroon. – Holotype (by orig. design.): MHNG 1010-52.

Distribution: Manengouba Mts., Cameroon.

Note added in proof:

Rhampholeon uluguruensis TILBURY & EMMRICH

1996 *Rhampholeon uluguruensis* TILBURY & EMMRICH, Trop. Zool., **9**: 62; figs 1 & 2. – Terra typica: Bondwa Peak area (1650 m alt.), Uluguru Mountains, Morogoro District, Tanzania. – Holotype (by orig. design.): ZMB 48421.

Distribution: Known only from the type locality.

Brookesia exarmata SCHIMMENTI & JESU

1996 *Brookesia exarmata* SCHIMMENTI & JESU, Ital. J. Zool., **63**: 194; figs 1–3. – Terra typica: Ambodyreana River, Tsingy de Bemaraha, Antsalova, Fivondronana, Mahajanga Province, Madagascar. – Holotype (by orig. design.): MRSN R1557.

Distribution: Known only from the type locality.

Comment: In their paper on *B. exarmata* SCHIMMENTI & JESU (1996: 194, 196) also studied related taxa, viz. *B. minima*, *B. peyrierasi*, *B. tuberculata*, *B. ramanantsoai*, and they (re-) established the morphological distinction between the first three taxa mentioned. Consequently *B. peyrierasi* and *B. tuberculata* were resurrected from the synonymy of *B. minima* (*contra* RAXWORTHY & NUSSBAUM, 1995), thereby confirming our reservations as to the proposed synonymy as indicated in the comment on *B. minima* in the Appendix of the present paper. No morphological differences between *B. dentata* and *B. ramanantsoai* were found, so the synonymisation of the latter form with the former remains valid.

Appendix

Comments on problems of taxonomy, typology, nomenclature, authorship etc. marked in the text

Chamaeleonidae

According to art. 29 (a) of the ICZN family names are formed by adding the latinized suffix – idae to the stem of the name of the type genus. The family name Chamaeleontidae (often used in American literature) is based on the genus name *Chamaeleon*, the ending of which derives from the Greek leon, genitive leontos = lion, hence the stem is leont –. However, as the genus name *Chamaeleon* GRONOVIUS (1763) constitutes a nomen rejectum (Opinion 89), this family name is to be rejected as well. The valid name of the type genus is *Chamaeleo*, the ending of which derives from the Latin leo, genitive leonis = lion. The stem as derived from the genitive is leon-, hence the proper family name is Chamaeleonidae. Even if the ending of the name of the type genus is considered a Greek word latinized with a change in termination, i.e. leo instead of leon, the stem is that appropriate to the latinized form, viz. leon-, which also results in the family name Chamaeleonidae (cf. art. 29 (b) (iii).

Brookesia GRAY

The original description of *Brookesia* by GRAY appeared in the volume of the Proceedings of the zoological Society of London of 1864, that was, however, published in May 1865 (DUNCAN 1937: 72).

LOVERIDGE (1956: 2, footnote; and 1957a: 206, footnote) designated *Brookesia nasus* as type species of *Evoluticauda*, but this subsequent designation is not valid as the subsequent designation of *Brookesia tuberculata* as type species of *Evoluticauda* by GUIBÉ (1954: 108) has precedence.

Brookesia decaryi ANGEL

ANGEL (1939: 577 and 1942: 182) mentioned five syntypes with the collection numbers MHNP 1938. 153–157. BRYGOO (1978: 16) repeated the indication of ANGEL (1939, 1942). GUIBÉ (1954: 108) gave the collection numbers [19]38. 158–162 for five syntypes, and BRYGOO (1983: 9) mentioned the collection numbers to be MHNP [19]38. 158–160 and [19]38. 161–162 as according to him ANGEL (1939, 1942) erred with respect to the collection numbers.

Brookesia dentata MOCQUARD

RAXWORTHY & NUSSBAUM (1955: 530) synonymized *B. ramanantsoai* with *B. dentata* on the basis of the joint possession of two pairs of laterally projecting tubercles in the anterior half of the body. Their series consisted of six specimens, including the types of *B. dentata* and *B. ramanantsoai.* A seventh specimen (ZFMK 8898), a ♂ from Périnet (BÖHME 1974: 80; identified as *B. ramanantsoai* by BRYGOO 1978: 69) has three pairs of tubercles, thus refuting the key of RAXWORTHY & NUSSBAUM (l.c.: 548). In view of the small-scaled endemism of *Brookesia* in northern Madagascar (see RAXWORTHY & NUSSBAUM, op.cit.) the lumping of extremely small samples from geographically widely isolated areas (among them the well-known centre of endemicity Périnet) seems premature.

Brookesia ebenaui (BOETTGER)

In his original description of *Brookesia ebenaui*, BOETTGER (1880a: 280) explicitly mentioned one specimen collected by Ebenau on which he based his description. In his typelist BOETTGER (1893a: IX) referred to the original description and gave the collection number, viz.: '6475a' under which the specimen was registered. BOETTGER (1893a: 124) also mentioned in the catalogue part of his publication two adult specimens of *B. ebenaui* to be present in the collection under number 6475a, one donated by Ebenau in 1880, the other donated by Stumpff in 1883. MERTENS (1922: 177) mentioned in his typelist under the subheading *B. ebenaui* the collection number '6475, a'. The comma presumably indicating that two type specimens were thought to be present (6475 and 6475a), an observation that is corroborated by his reference to the donators Ebenau and Stumpff. Both donations are referred to have taken place in 1880. In his second typelist, MERTENS (1967: 65) then considered the two specimens listed by BOETTGER (1893a) and himself in 1922 to be syntypes and subsequently designated a lectotype: 'SMF 16516 (= Kat. BOETTGER 6475a)'. He repeated the donation of the two specimens by Ebenau and Stumpff to have taken place in 1880. In view of all this we consider the various indications and the subsequent designation of a lectotype by MERTENS (1922, 1967) to be erroneous and the single specimen that was donated by Ebenau in 1880 and on which BOETTGER (1880a) based his original description to be the holotype (by monotypy). This observation is corroborated by BOETTGER (1884: 61) who mentioned the donation of the „zweite überhaupt bekannte Exemplar von *Chamaeleo ebenaui* BTTG." by Stumpff during the period 1883–1884. BRYGOO (1978: 28) also mentioned one type specimen.

RAXWORTHY & NUSSBAUM (1995: 532) were not able to study the holotype of *B. ebenaui* and based their synonymization on published figures and descriptions.

We had the opportunity to examine the holotype (SMF 16515) and we can confirm the conclusions of RAXWORTHY & NUSSBAUM.

Brookesia minima BOETTGER

As to the synonymization of *B. tuberculata* and *B. peyrierasi* with *B. minima* (RAXOWRTHY & NUSSBAUM 1995: 527f.) we are of the same opinion as expressed regarding the synonymization of *B. ramanantsoai* with *B. dentata* (see above). The authors do not consider the hemipenial differences between *B. tuberculata* and *B. peyrierasi* as described by BRYGOO & DOMERGUE (1969, 1975). Moreover, our material of topotypical *B. minima* and *B. peyrierasi* (ZFMK) shows a striking difference in body size between the two populations.

Brookesia stumpffi BOETTGER

BOETTGER (1894: 183, 184) in his original description of *B. stumpffi*, gave the measurements of one ♂ and one ♀, but mentioned numerous specimens at his disposal from Nossibe, collected by Ebenau and Stumpff and donated to the Frankfurt museum, and one ♂ specimen from Soalala, W. Madagascar collected by Voeltzkow. The specimens were registered under the collection number 6474a–b. Whether the one ♂ specimen collected by Voeltzkow was included in the type series present in the Frankfurt collection is not known. MERTENS (1922: 177) referred to the original description of BOETTGER (1894), but gave the collection number '6474, a'. Moreover, he mentioned only Nossibé as type locality and referred only to Ebenau and Stumpff as donators of the type material, thus neglecting the ♂ specimen from Soalala collected by Voeltzkow. MERTENS (1967: 64) surprisingly regarded 'SMF 16475 (Kat. MERTENS 6474a (sic))' to be the holotype of *B. stumpffi*. As the original description was based on at least three specimens, but probably more, these specimens are to be considered to constitute a syntype series with the terra typica Nossibe and Soalala. The various indications by MERTENS (1922, 1967) concerning collection number, type locality, collectors and especially the putative holotype are to be considered erroneous. BRYGOO (1978: 71) confirmed this view as he mentioned numerous type specimens collected by Ebenau, Stumpff and Voeltzkow and the type locality as Nosy Bé and Soalala. The precise composition of the syntype series is unknown at present, but the series consists at least of all specimens referred to by BOETTGER (1894) in his original description, whereever they have been deposited. In order to avoid further confusion we designate SMF 16475 as lectotype with the consequence that the type locality becomes Nosy Bé. The remaining specimens of the syntype series (including the ♂ specimen from Soalala) thereby become paralectotypes.

Brookesia superciliaris (KUHL)

KUHL (1820: 103) mentioned the type(s) of *B. superciliaris* to be present in the BMNH, but the type(s) is/are not registered as such (C. McCarthy, in litt. 1-II-1995).

Rhampholeon boulengeri STEINDACHNER

In view of unpublished observations on the hemipenes of *R. s. spectrum* and *R. s. boulengeri*, it is certain that these taxa cannot be conspecific. The hemipenes of *R. boulengeri* have an apical ornamentation consisting of papillate horns, comparable to that of the majority of other Central and East African *Rhampholeon* species (KLAVER & BÖHME 1986). The apical ornamentation of the hemipenes of *R. spectrum* differs profoundly from that of *R. boulengeri* and conforms the tentative description by KLAVER & BÖHME (1986: 32). Consequently we treat *boulengeri* as a valid species.

Rhampholeon chapmanorum TILBURY

In his original description TILBURY (1992: 9) mentioned that he named the new chameleon after JIM and BETTY CHAPMAN. According to article 31 (a) (ii) of the International Code of Zoological Nomenclature the new name is to be formed by adding *-orum* to the personal name „if [the personal name is that] of men or of man (men) and woman (women) together, ...", whereas *-i* is added „if the personal name is that of a man, ...". Consequently the name *chapmani* constitutes an incorrect original spelling that has to be corrected (cf. art. 32 (c) and (d) of the ICZN), hence *chapmanorum* instead of *chapmani*.

Rhampholeon spectrum (BUCHHOLZ)

The original description of BUCHHOLZ appeared in the volume of the „Monatsberichte" of 1874, but from the title-page it becomes clear that this volume was published in 1875 (but see also KLAVER & BÖHME 1992: 455).

Although BUCHHOLZ (1875: 82) referred to „eine größere Anzahl von Exemplaren" he collected, only two syntypes were traced in the collection of the ZMB.

Although *R. affinis* is considered synonymous with *R. boulengeri* (DE WITTE 1965: 192), we treat it tentatively as synonymous with *R. spectrum* in view of the apical ornamentation of its hemipenes (BÖHME, unpubl. observ.) that is quite different from that of *R. boulengeri* and very similar to that of *R. spectrum*. In view of the vastly disjunct geographical distribution and the observations of SCHMIDT (1919) and LOVERIDGE (1942) it is likely that *R. affinis* is to be revalidated as a species.

Calumma GRAY

The original description of *Calumma, Crassonota* and *Archaius* by GRAY appeared in the volume of the Proceedings of 1864, that was, however, published in May 1865 (DUNCAN 1937: 72).

Calumma brevicornis hilleniusi (BRYGOO, BLANC & DOMERGUE)

In view of the striking size difference and differences in hemipenial morphology, *hilleniusi* might prove to be a distinct species. This is also implied by the hierarchy of differentation level if the subspecific status of *B. b. tsarafidyi* is considered.

Calumma brevicornis tsarafidyi (BRYGOO & DOMERGUE)

BRYGOO (1983: 19) indicated the specimen MHNP A172 to be holotype (by monotypy). However, although the original description was based on a single specimen, this specimen was explicitly referred to as 'le type' and therefore has to be considered holotype (by original designation).

Calumma gastrotaenia (BOULENGER)

The profound differences in hemipenes ornamentation between the various subspecies of *C. gastrotaenia* may be an indication that the status of these taxa has to be reappraised in future research (see also KLAVER & BÖHME 1986: 34).

Calumma oshaughnessyi (GÜNTHER)

C. oshaughnessyi was originally described as *Chamaeleon o'shaughnessii.* As the species was named after its collector ARTHUR O'SHAUGHNESSY, the original spelling of the name is clearly an incorrect original spelling that is to be corrected according art. 32 (d) of the Code of Nomenclature. The proper name is *C. oshaughnessyi* as was realised by all subsequent authors who used this name.

Calumma tsaratananensis (BRYGOO & DOMERGUE)

BRYGOO (1983: 20) indicated the specimen MHNP A59 to be holotype (by monotype). However, although the original description was based on a single specimen, this specimen was explicitly referred to as 'le type' and therefore has to be considered holotype (by original designation).

Furcifer FITZINGER

The original description of *Apola, Sauroceras, Dicranosaura* and *Cyneosaura* by GRAY appeared in the volume of the Proceedings of 1864, that was, however, published in May 1865 (DUNCAN 1937: 72).

Furcifer angeli (BRYGOO & DOMERGUE)

BRYGOO (1983: 5) indicated the specimen MHNP 1967. 174 to be holotype (by monotypy). However, although the original description was based on a single specimen, this specimen was explicitly referred to as 'le type' and therefore has to be considered holotype (by original designation).

Furcifer balteatus (Duméril & Bibron)

With regards to the authorship of the name *F. balteatus* see comment adjoining *Chamaeleo calyptratus*.

Furcifer belalandaensis (Brygoo & Domergue)

Brygoo (1983: 7) indicated the specimen MHNP 1969. 114) to be holotype (by monotypy). However, although the original description was based on a single specimen, this specimen was explicitly referred to as 'le type' and therefore has to be considered holotype (by original designation).

Furcifer lateralis (Gray)

Brygoo (1971: 164) discussed the morphological differences between a *lateralis* population from S.W. Madagascar and the typical (and sympatric) *lateralis* population. Specimens of this particular population are characterized by their larger size and their aberrant scalation and colouration. As Brygoo indicated, this population as „forme *major*" (his italics) and even mentioned a particular specimen (MHNP 457/C) and a specific locality (Tanandava), we consider *major* a formal synonym of *lateralis* and note that it may eventually be elevated to a (sub-)specific rank. Brygoo (1978: 141) also mentioned *major* from various other localities.

Furcifer oustaleti (Mocquard)

In his original description, Mocquard (1894: 3) referred to numerous specimens, viz.: 'un de Betsileo par Catat, et six de Diégo Suarez par Alluaud & Belly', so at least 7 specimens appeared to have been at his disposal at the time. Angel (1942: 164) registered 7 specimens as well with the collection numbers 91–22, 91.323–324 and 93.155–158. Guibé (1954: 107) referred to 1 specimen from Betsileo and only 5 specimens from Diégo Suarez, resulting in 6 syntypes registered under 5 collection numbers: 91.22 and 91.155–158 respectively. Brygoo (1971: 201) repeated this indication of the six syntypes, but Brygoo (1983: 15) referred to only 5 syntypes, 4 from Diégo Suarez and 1 from Betsileo. No explanation as to this dwindling number of syntypes can be given, so we follow Brygoo (1983) and consider the remaining syntype series to consist of 5 specimens.

Both *F. oustaleti* and *F. verrucosus* exhibit hemipenial and karyotypical variability that might be an indication that more than one taxon is covered by the respective names (Klaver & Böhme 1986: 31; footnote and table 1).

Furcifer pardalis (Cuvier)

Werner (1899: 184) mentioned two ♀ specimens of *Ch. axillaris* to be present in „Mus. Berolin.", however, these specimens could not be found in the ZMB (R. Günther, in litt. 23.III.1995).

Furcifer verrucosus (CUVIER)

See comment on *F. oustaleti*

Bradypodion FITZINGER

The original description of *Microsaura* and *Lophosaura* by GRAY appeared in the volume of the Proceedings of 1864, that was, however, published in May 1865 (DUNCAN 1937: 72).

Bradypodion fischeri (REICHENOW)

MERTENS (1966: 13–15) adopted LOVERIDGE's (1957a: 203–204) classification and considered the taxa *excubitor, multituberculatum, tavetanum, uluguruense* and *uthmoelleri* subspecies of *B. fischeri*, which resulted in a quite variable 'species' with a vastly disjunct distribution. The argumentation of LOVERIDGE was not very substantial, and it is not surprising that *tavetanum* and *uthmoelleri* are presently considered valid species again (fide BROADLEY & HOWELL 1991 and BÖHME & KLAVER 1990). Of the remaining taxa, *excubitor* stands out from the *fischeri*-like forms in external morphology and its isolated geographical distribution. Indeed *excubitor* was originally described as a subspecies of *B. tenue* with which, according to its original author, it has much in common, but BARBOUR & LOVERIDGE (1928: 176) simply stated it to represent a race of *B. fischeri*. Nevertheless, we treat *excubitor* as well as *multituberculatum* and *uluguruense* as subspecies of *B. fischeri*, but note that the necessary revision of these taxa is likely to lead to a reevaluation of their rank (see also EMMRICH 1994).

Bradypodion pumilum (GMELIN)

The status of *B. pumilum* and allies remains to the present day a matter in dispute. The taxa concerned, viz.: *pumilum, caffer, damaranum, gutturale, karrooicum, melanocephalum, occidentale, taeniabronchum, transvaalense* and *ventrale* were studied in detail by METHUEN & HEWITT (1914), POWER (1932), HEWITT (1935), FITZSIMONS (1943) and HILLENIUS (1959). Two major questions were addressed to in these studies: 1) the generic status of this group of taxa and 2) the status of the taxa themselves. METHUEN & HEWITT (1914: 103) were, apart from GRAY (1865), the first to propose a separate generic status for *B. pumilum* and allies and restored the name *Lophosaura* GRAY 1865. This generic name was adopted by POWER (1932) and HEWITT (1935), but FITZSIMONS (1943) pointed out that *Lophosaura* was a junior synonym of *Microsaura* GRAY 1865. Subsequently, LOVERIDGE (1957a: 197) and KLAVER (1973: 156) indicated that *Microsaura* was in turn a junior synonym of *Bradypodion* FITZINGER 1843. RAW (1976) was the first South African author to finally use the correct generic name when describing three new taxa of the *pumilum* group, whereas KLAVER & BÖHME (1986) included next to *B. pumilum* and allies various central and east African species in *Bradypodion*. Criticism as to the justification of the genus concept *sensu* RAW (1976) was voiced by KLAVER (1981: 48) and that *sensu* KLAVER & BÖHME (1986), by HOFMAN & al. (1991: 260). Nevertheless, as was already indicated in the introduction, we adopt in this list the genus concept as proposed by KLAVER & BÖHME (1986) that, despite its shortcomings, appears to be a better concept than that which lumps all ordinary chameleons in one genus *Chamaeleo* (fide MERTENS 1966).

The history of the status of the taxa of the *pumilum* group is fraught with confusion as a result of misidentifications, overlapping variation in characters, uncertainties in geographical distribution etc. HILLENIUS (1959), the last author to study the taxa in detail and to try to sort out the confusion, concluded the taxa to be subspecific forms of *B. pumilum*. His conclusion was subsequently adopted by MERTENS (1966) and BURRAGE (1972). RAW (1976 & 1978), however, regarded them again as valid species, but his argumentation to that end is wanting as a result of his virtually ignoring of the contents and conclusions of the previous studies and his urge to start afresh by adding four new taxon names to the fray. He was, uncritically, followed by KLAVER & BÖHME (1986) and BRANCH (1988). HOFMAN & al. (1991) then regarded several of the new species to be subspecific forms of *B. pumilum*, entirely in vein of HILLENIUS' original reasoning, but quite unjustified. BRANCH (1988: 180–181) referred to the problematic status of *B. pumilum* and allies and thought the argumentation as to a specific as well as a subspecific status wanting. He even mentioned *B. gutturale* and *B. karrooicum* to be best considered subspecies of *B. ventrale*, but surprisingly treated most of the taxa involved as valid species. As a consequence the matter as to the status remains unresolved until a comprehensive revision of available literature and collection material is executed. In view of this we adopt in this list the most recent critical literature, i.e. HILLENIUS (1959), and treat the taxa concerned as subspecies of *B. pumilum*. The species described by RAW (1976, 1978), viz.: *dracomontanum, nemorale, setaroi* and *thamnobates*, are, of course, treated as valid species.

DUMÉRIL & BIBRON (1836: 219) synonymized *Chamaeleo bonae spei* LAURENTI, 1768 with *Chamaeleo senegalensis* as they considered the illustration its description was based on (SEBA, 1734, I; plate 83, fig. 5) to be a poor representation of the latter species (see also WERNER, 1911: 12 and MERTENS, 1966: 27). However poor the illustration may be, the adjoining description of the casque, the colouration and the scalation (as well as the place of origin as deduced from its name) suggests another possibility, i.e. that *B. pumilum* is involved. This assumption was quite unexpectedly supported by our recent discovery of a publication in which a (completely forgotten) chameleon species was described (GRAVENHORST, 1807: 429). Considering the description of the head, the gular crest, the body scalation and the colouration of *C[hamaeleo] thermophilus*, it is most likely that *B. pumilum* (possibly *damaranum*) is involved again. Moreover, in the description a reference is made to what GRAVENHORST considered comparable specimens, viz. the illustrations of SEBA (1734; plate 82, fig. 4 and plate 83, fig. 5). DUMÉRIL & BIBRON (1836: 217) already considered SEBA's plate 82, figs. 4 & 5 possible representations of *B. pumilum*. In view of these combined observations, we consider *Chamaeleo bonae spei* not a likely synonym of *Ch. senegalensis*, but of *B. pumilum*. The synonymization of *Ch. thermophilus* with *Ch. bonae spei* and also with *B. pumilum* is on a more secure footing which makes the synonymization of *Ch. bonae spei* with *B. pumilum* even more credible.

Bradypodion pumilum gutturale (SMITH)

FITZSIMONS (1935/36: 270) could not locate any type(s) of *B. gutturale*, and FITZSIMONS (1943: 168) presumed the type(s) to be lost. However, examination of the partially undetermined chameleon collection of A. SMITH present at the NMS (no. 1859.13) is desirable to verify this opinion.

Subgenus Chamaeleo LAURENTI

The original description of *Phumanola, Calyptrosaura, Erizia* and *Dilepis* by GRAY appeared in the volume of the Proceedings of 1864, that was, however, published in May 1865 (DUNCAN 1937: 72).

Chamaeleo (Ch.) africanus LAURENTI

The type(s) of *Ch. candidus* is/are present in „Museo Turriano", which was the private collection of Count „Francisci Annibalis Turriani etc. etc." (LAURENTI 1768: 26). This collection could not be located and, to be sure, is not at the museum in Torino (Turin), i.e. MRSN (F. Andreone, in litt. 4-II-1995) nor at the NMW.

Chamaeleo (Ch.) anchietae BOCAGE

Despite the synonymy of *mertensi, vinckei* and *marunguensis* with *anchietae*, demonstrated by LOVERIDGE (1957a) and HILLENIUS (1959), MERTENS (1966) treated them as valid subspecies without any argumentation. We adopt the view of the first two authors mentioned.

Chamaeleo (Ch.) calyptratus DUMÉRIL & BIBRON

The history of the authorship of *Ch. calyptratus* and *F. balteatus* is chameleon-like, partly because of varying opinions as to whom the original description is to be attributed and partly because of varying opinions as to whom the original publication is to be attributed. First, the publication in which both species were described is the well-known 'Catalogue méthodique' of 1851 and its title-page is quite unequivocal as to its authors: 'M. C. DUMÉRIL' and 'M. AUG. DUMÉRIL'. That would have ended the matter before it had started were it not for numerous quotations of A[UGUST] DUMÉRIL as sole author of the Catalogue, e.g. WERNER

(1902), MERTENS (1966) & al. The reason behind this is that AUGUST DUMÉRIL allegedly had written the text of the Catalogue all by himself. This may have been the case, but what matters to the bibliographer is that the text was published under the name of his father and himself and this leaves no doubt whatsoever about the formal authorship of the publication.

The problem of the authorship of the names *Ch. calyptratus* and *F. balteatus* is rather more complicated. Again the original authorship seems quite unproblematic as both new names are followed by the indication of its authors, viz.: ‚DUM. BIB.'. 'DUM.' most probably stands for CONSTANT DUMÉRIL (father of AUGUST) and 'BIB.' for GABRIEL BIBRON, both authors of the renowned „Erpétologie générale". Subsequent citations of the two species names reveal, however, the following indications of authorship:

1) A. DUMÉRIL (see DUMÉRIL, 1852; WERNER, 1902; GUIBÉ, 1954, MERTENS, 1966);

2) A. & C. (or C. & A.) DUMÉRIL (see WERNER, 1911; HILLENIUS, 1966 & 1978); BRYGOO, 1971 (*calyptratus*), 1978 & 1983) and

3) DUMÉRIL & BIBRON (see BRYGOO, 1971 (*balteatus*) & 1978; BRYGOO & DOMERGUE, 1969).

An explanation that A. DUMÉRIL is considered the author of *Ch. calyptratus* and *F. balteatus* might be that he is considered the sole author of the Catalogue, but this is not tenable in view what has been argued above . Incidentally, in his publication of 1852. A. DUMÉRIL added to the confusion by writing after the names of the species concerned „A. DUMÉRIL, espèce nouvelle". Similarly, when C. & A. DUMÉRIL are considered the authors of the Catalogue they might, by inference, be considered the authors of the new species names, although art. 50(a) of the Code still allows the possibility only A. DUMÉRIL (assuming to have written the text of the Catalogue singlehandedly or at least the original descriptions) to be attributed with the authorship of the two species names. In both cases, however, an intentional negligence of the indication „DUM. BIB." has to be accounted for. Why should A. DUMÉRIL, assuming him to be the sole author of the text of the Catalogue, have mentioned 'DUM. BIB.' at all? An explanation might be that, after succeeding the deceased GABRIEL BIBRON at the Paris museum, he built on the manuscripts of C. DUMÉRIL and G. BIBRON and using these for the composition of the text of the Catalogue paid duely hommage to its original authors, hence „DUM. BIB.". In any case, and apart from any opinion one may have of all matters mentioned above, art. 50(a) (i) of the Code also allows on the basis of the indication „DUM. BIB." that [C.] DUMÉRIL & BIBRON be considered the authors of the names *Ch. calyptratus* and *F. balteatus*. Apart from his previous inconsistent opinions of the matter, we think the argumentation of BRYGOO (1983: 6 & 8) inadequate and, therefore, his final and consistent opinion as to authorship of the two names untenable. Consequently, we think it justified to consider. [C.] DUMÉRIL & BIBRON the authors of the names *Ch. calyptratus* and *F. balteatus*.

The *terra typica* of *Ch. calyptratus* is also a matter of misunderstanding, the cause of which can be found in the „Catalogue méthodique". On page 31 DUMÉRIL & BIBRON described *Ch. calyptratus* without any indication of a *terra typica*, but at the bottom of page 33 it is stated: „N.B. Le C. à cape [= *calyptratus*] est orig. de l'Afri. (rég. du Nil) ...". Subsequent authors failed to notice this which led to erroneous references, e. g. MERTENS (1966: 9) and BRYGOO (1971: 276). BRYGOO (1983: 8) corrected himself, but failed to do justice to HILLENIUS (1959: 47, 1966: 91), who indicated correctly the *terra typica*, but was criticized for this by BRYGOO (1971).

Chamaeleo (Ch.) calyptratus calcarifer PETERS

In his first paper on Arabian chameleons, HILLENIUS (1966: 103) depicted the type of *calcarifer* and hypothesized it to be a hybrid of *Ch. calyptratus* and *Ch. arabicus*. Consequently it must originate from an area of overlap of the distribution of these two species (i.e. Yemen). Later HILLENIUS & GASPERETTI (1984: 520) referred the name *calcarifer* to a northern, viz.: Saudi Arabian population of *Ch. calyptratus*, that has a distinctly different appearance. Laboratory bred hybrids of *Ch. calyptratus* and *Ch. arabicus*, deposited in the ZFMK, match the external morphology of the ZMB type specimen of *calcarifer* very well. Therefore, we think that HILLENIUS' original assumption as to the hybrid nature of *calcarifer* was correct and the subsequent application of this name to indicate the Saudi Arabian *calyptratus* population incorrect. If future studies will prove this to be the case, then this form of *calyptratus* has to be renamed.

Chamaeleo (Ch.) chamaeleon (LINNAEUS)

If the type locality of *Ch. fasciatus* is correct, then this name is most likely not a synonym of *Ch. chamaeleon*, but of *Ch. gracilis*. Unfortunately, this could not be verified as the type specimen(s) could not be found in the collection of J. A. Smith present at the NMS (G. Swinney, in litt. 24. V. 1995).

Chamaeleo (Ch.) chamaeleon chamaeleon (LINNAEUS)

The restriction of the type locality of *Ch. cinereus* to 'Argélia' by CRESPO (1972: 475) has no standing as the restriction of MERTENS & WERMUTH (1960: 85) has precedence. BOETTGER (1880b: 198) described *Chamaeleo vulgaris* var. *recticrista* on the basis of 15 specimens: „2 junge Stücke (no. 5 und 6) von Haiffa (F. Lange), 4 weitere (No. 7–10) aus der Umgebung von Jerusalem aus der H. Simon'schen Schenkung, 4 Stücke von Beyrut und 5 Stücke von Cypern (W. Schlüter) ...". He gave also the measurements of no. 5–10 and concluded on p. 199: „... die syrisch-cyprisch-palästinensische Form mit einem besonderen

Namen (*recticrista*) zu bezeichnen." In his Katalog BOETTGER (1893a: IX) did not refer to the types of *recticrista* in his typelist, but in the catalogue part of his publication mentioned: „6443 h ♀ (var. *recticrista* BTTGR). Smyrna, Gesch. 1870 von Hofrat Dr.med. Pauli, hier." Besides, he mentioned „6443 f. 6 jung. Jerusalem. Gesch. 1880 von Hans Simon, Stuttgart" and „6443 g. 3 jung. Haiffa. Gesch. 1879 von demselben." without any further indication and did not mention any material from Beyrut and Cyprus at all. MERTENS (1922: 177) in his typelist referred to what he considered the type(s): „Jerusalem. Nr. 6443, f. H. Simon G. 1880." and MERTENS (1967: 65) designated a „Lectotypus SMF 16151 (= Kat. Boettger 6443 f.). ♂, s. ad. Jerusalem; H. Simon l. d. 1880." Although all this is rather confusing as probably both BOETTGER (1893a) and MERTENS (1922) erred, MERTENS (1967) seems to have selected a specimen from the original syntype series and, therefore, the designation of the lectotype SMF 16151 is valid.

Although BOULENGER (1887: 443), WERNER (1911: 10) and MERTENS (1922: 177) already considered *Ch. vulgaris recticrista* synonymous with *Ch. ch. chamaeleon*, MERTENS (1966: 11) surprisingly treated *recticrista* as a valid subspecies. HILLENIUS (1978: 54) put the matter to rest again. Nevertheless, we have some reservations as to the synonymization of *recticrista* with *Ch. ch. chamaeleon*. Data from Turkey and Greece suggest that eastern Mediterranean representatives of *Ch. ch. chamaeleon* attain much larger dimensions than western Mediterranean ones (from Portugal and Spain). Furthermore, hemipenial differences between representatives of both groups and their separation by another distinct form, viz. *Ch. ch. musae* may also indicate a taxonomic distinction (BÖHME & WIEDL 1994: 33).

DUMÉRIL & BIBRON (1836: 205), ARADAS (1868: 440) and MERTENS (1966: 10) referred to „*Cameleo siculus* GROHMANN, 1832". However, in his article in the „Giornale di scienze, lettere e arti per la Sicilia, Palermo, 39: 145–153, GROHMANN only used the vernacular name „Camaleonte siculo". Therefore we consider the name employed by DUMÉRIL & BIBRON, ARADAS and MERTENS a nomen nudum.

Chamaeleo (Ch.) dilepis LEACH

A revision of *Ch. dilepis* and its subspecies is a challenge that nobody has yet dared to venture upon. Regrettable but understandable in view of the necessary study of the vast amount of material present in museum collections to be able to assess the subtle, if uninspiring, variation in distinguishing characters. As a consequence we shall have to accept the dubious classification as it is. It is likely that some taxa, e. g. *roperi*, might prove to be valid species, whereas other taxa might not even be considered valid subspecies and become synonyms.

Chamaeleo (Ch.) dilepis ruspolii BOETTGER

BOETTGER (1893b: 117) based his original description of *Ch. ruspolii* on one ♂ and one ♀ specimen. MERTENS (1922: 179) gave in his typelist as collection number „6450, 2a" suggesting more than one type specimen, but surprisingly MERTENS (1967: 65) mentioned „SMF 16300 (= Kat. Mertens 6450, 2a). ♂,..." to be the holotype without any argumentation. Of course, this ♂ specimen cannot be regarded as a holotype, and to avoid further confusion we designate here SMF 16300 as lectotype.

Chamaeleo (Ch.) laevigatus GRAY

MONARD (1951: 141) recorded four syntypes of *tibatiensis* with the (field) numbers 767, 775, 789 and 832. Number 789 is presumed lost, whereas the remaining three specimens are presently registered as MHNC 91.063–065 (M. JACQUAT, pers. comm.). An additional specimen (field number 832a, now MHNC 91.062) has later been added to the series and has subsequently been labelled as holotype. However, as this specimen was not mentioned by MONARD in his original description, it does not form part of the syntypic series and even if it did, it cannot be regarded as holotype. The three remaining original specimens match the measurements given by MONARD (l.c.), but after examination MHNC 91.065 proved to be a juvenile *Ch. gracilis*. Therefore, a lectotype designation is even more necessary. We designate here MHNC 91.063 as lectotype of *tibatiensis*, MHNC 91.064–065 thus becoming paralectotypes.

Chamaeleo (Ch.) namaquensis SMITH

According to FITZSIMONS (1935/36: 270 & 1943: 157) specimen(s) of *Ch. namaquensis* present in the BMNH might be regarded as type(s). He also mentioned that original material of A. SMITH was presented to the MHNP and the Royal Scottish Museum (presently the NMS) in Edinburgh. In the collection of the latter FITZSIMONS (1935/36) did not find any types of *Ch. namaquensis*, but it remains to be seen if in the partially undetermined collection of chameleons of A. SMITH present at Edinburgh (NMS 1859.13) there might be a type specimen present. BRYGOO (1983: 14) reported on the presence of one of the types at the MHNP. At the BMNH no types of *Ch. namaquensis* are recognized (C. McCarthy, in litt. 1.II.1995).

Subgenus Trioceros SWAINSON

The original description of *Pterosaurus* and *Ensirostris* by GRAY appeared in the volume of the Proceedings of 1864, that was, however, published in May 1865 (DUNCAN 1937: 72).

Chamaeleo (T.) bitaeniatus FISCHER

HELLMICH (1956: 119) recorded *Ch. bitaeniatus* from Litembo, between Lake Nyasa and Kigonsera, S. Tanzania, but according to RAND (1963: 20) this record is likely to concern *Ch. rudis*.

Chamaeleo (T.) chapini DE WITTE

According to the „Names of persons: national usages for entry in catalogues" (IFLA, 1977) Belgian surnames with a prefix are entered under the first letter of the prefix. We adopt here these internationally accepted rules, hence „DE WITTE" and not „WITTE, DE".

Chamaeleo (T.) ituriensis SCHMIDT

MERTENS (1966) adopted the point of view of LOVERIDGE (1942), i.e. considered *ituriensis* without argumentation as a subspecies of *Ch. johnstoni*. We follow SCHMIDT (1919) and DE WITTE (1965) and treat *ituriensis* as a valid species.

Chamaeleo (T.) jacksonii jacksonii BOULENGER

EASON & al. (1988: 581, 586) recorded a male specimen of *Ch. jacksonii* from Mombo, Usambara Mts., Tanzania, but were unsure if the specimen belonged to *Ch. j. jacksonii*. As *Ch. jacksonii* was never recorded from this locality before or since, we think it more likely that either the locality record is erroneous or the determination is (e.g. confounded with the three horned *Ch. deremensis* from the Usambara Mts. (see also BROADLEY & HOWELL 1991: 13).

Chamaeleo (T.) jacksonii xantholophus EASON, FERGUSON & HEBRARD

EASON & al. (1988: 587) assumed one male specimen from Mt. Marsabit they studied to belong to *Ch. j. xantholophus*. TILBURY (1991: 164) described *Ch. marsabitensis* and assumed a sympatric distribution of the two forms on Mt. Marsabit. In view of the variation in characters and the small sample size (one specimen) in the study of EASON & al. it is in our opinion more likely that the putative *Ch. j. xantholophus* is not sympatric but conspecific with *Ch. marsabitensis*. Moreover, in view of the limited distribution of *Ch. j. jacksonii* and *Ch. j. xantholophus* on Mt. Kenya, the validity of the latter seems doubtful.

Chamaeleo (T.) johnstoni BOULENGER

HELLMICH (1956: 122) recorded *Ch. j. johnstoni* from Peramiho, West of Songea, S. Tanzania. In view of the geographical distribution of *Ch. johnstoni* as it is presently known and the fact that only a single juvenile specimen was concerned we question the validity of this determination.

Chamaeleo (T.) rudis BOULENGER

ZFMK material of *sternfeldi* from Kilimanjaro shows strikingly different dimensions and body proportions as compared with *rudis* from the Ruwenzori

(BÖHME, unpubl.). In addition, colour photographs of *sternfeldi*, deposited in the ZFMK as well, show striking differences in colour pattern as compared with *rudis*. Consequently we think this may be an indication that *sternfeldi* is a valid species.

Chamaeleo (T.) tempeli TORNIER

In the introduction of his article, TORNIER (1900: 58) acknowledged the people who put the specimens at his disposal, one of which was a missionary named „Wolf". On page 613 he mentioned the missionary again, but now by the name „Wolff". Subsequently on page 614 he named the new taxon of *Ch. tempeli* „var. *wolfi*", but the legend of figure G reads „var. *wolffi*". TORNIER (1902: 590) referred again to „var. *wolffi*" and this name was adopted by WERNER (1911: 25) and MERTENS (1966: 28) as well. LOVERIDGE (1933a: 199) referred to „*wolffi*", but on page 336 he mentioned „*wolfi*" again. It is clear that in the article of TORNIER (1900) either „*wolfi*" or „*wolffi*" is a lapsus. As the article itself furnished no additional information, we adopt the view „*wolfi*" to be the incorrect original spelling and the subsequent spelling of TORNIER (1902, being the first reviser), WERNER (1911) and MERTENS (1966) the correct one.

HELLMICH (1956: 123) recorded one specimen of *Ch. tempeli* from 'Marangu (Kilimandscharo)'. The identification of the specimen is undoubtedly correct as it possesses the double V-shaped gular crest characteristic of *Ch. tempeli* (unless one assumes a new undescribed species with a similar trait). Whether the locality record is correct remains to be seen as *Ch. tempeli* has not been recorded from there before or since.

References

ANDERSSON, L.G. (1900): Catalogue of the Linnean type-specimens of Linnaeus's Reptilia in the royal museum in Stockholm. – Bih. K. Svenska Vet.-Akad. Handl., **26**: 3–29.

ANGEL, F. (1939): Lézards, Scincidés et Chamaeleontidés nouveaux de Madagascar. – Bull. Mus. Hist. nat., **10** (1938): 574–577.

– (1942): Les lézards de Madagascar. – Mém. Acad. malg., **36**: 1–193.

ARADAS, A. (1868): Sul Camaleonte siculo. – Atti Soc. ital. Sci. nat., **11**: 439–446.

BARBOUR, T. & LOVERIDGE, A. (1928): A comparative study of the herpetological faunae of the Uluguru and Usambara Mountains, Tanganyika Territory, with description of new species. – Mem. Mus. comp. Zool., **50**: 87–265.

BÖHME, W. (1974): Ein seltenes Zwergchamäleon aus Madagaskar. – Salamandra, **10**: 80–82.

– (1987): Seltene Reptilien aus dem Uzungwe-Gebirge, Süd-Tansania. – Herpetofauna, **9**: 27–34.

– & KLAVER, CH. (1981): Zur innerartlichen Gliederung und zur Artgeschichte von *Chamaeleo quadricornis* TORNIER, 1899 (Sauria: Chamaeleonidae). – Amphibia-Reptilia, **1**: 313–328.

– & – (1990): Zur Kenntnis von *Bradypodion uthmoelleri*. – Salamandra, **26**: 260–266.

– & WIEDL, H. (1994): Status and zoogeography of the herpetofauna of Cyprus, with taxonomic and natural history notes on selected species (genera *Rana, Coluber, Natrix, Vipera*). – Zool. Middle East, Heidelberg, **10**: 31–52.

BOETTGER, O. (1880a): Diagnoses reptilium et batrachiorum novorum a Carolo Ebenau in insula Nossi-Bé Madagascariensi lectorum. – Zool. Anz., **3**: 279–283.

– (1880b): Die Reptilien und Amphibien von Syrien, Palaestina und Cypern. – Ber. senckenb. naturf. Ges., **1880**: 132–219.

– (1884): Sectionsberichte. Herpetologische Section. – Ber. senckenb. naturf. Ges., **1884**: 61–62.

– (1893a): Katalog der Reptilien-Sammlung im Museum der senckenbergischen naturforschenden Gesellschaft in Frankfurt am Main. I. (Rhynchocephalen, Schildkröten, Krokodile, Eidechsen, Chamäleons). I–X, 1–140.

– (1893b): Übersicht der von Prof. C. KELLER anläßlich der RUSPOLI'schen Expedition nach den Somaliländern gesammelten Reptilien und Batrachier. – Zool. Anz., **16**: 113–119.

– (1894): Eine neue *Brookesia* (Chamaeleontidae) aus Nossibé. – Zool. Anz., **17**: 182–185.

BOULENGER, G.A. (1887): Catalogue of the lizards in the British Museum, **3**: 437–476.

– (1892): On newly-discovered East-African chameleons with remarks on some other reptiles described by Dr. STEINDACHNER. – Ann. Mag. nat. Hist., **9**: 72–74.

BOURGAT, R.M. (1970): Récherches écologiques et biologiques sur le *Chamaeleo pardalis* CUVIER 1829 de l'île de la Réunion et de Madagascar. – Bull. Soc. zool. France, **95**: 259–269.

BRANCH, B. (1988): Field guide to the snakes and other reptiles of southern Africa. – Struik, Cape Town, 328 pp.

BROADLEY, D.G. (1971): A review of *Rhampholeon marshalli* BOULENGER, with the description of a new subspecies from Mozambique (Sauria: Chamaeleonidae). – Arnoldia, **5**: 1–6.

– & HOWELL, K.M. (1991): A check list of the reptiles of Tanzania, with synoptic keys. – Syntarsus, **1**: 1–70.

BRYGOO, E.R. (1969): *Chamaeleo guentheri* BOULENGER 1888, synonyme de *Ch. pardalis* CUVIER 1829. – Bull. Mus. Hist. nat., **41**: 117–121.

– (1971): Reptiles Sauriens Chamaeleonidae. Genre *Chamaeleo*. – Faune de Madagascar, **33**: 1–318.

– (1972): Acquisitions récentes sur les caméléons de Madagascar. – Arch. Inst. Pasteur Madagascar, **41**: 143–155.

– (1978): Reptiles Sauriens Chamaeleonidae. Genre *Brookesia* et complément pour le genre *Chamaeleo*. – Faune de Madagascar, **47**: 1–173.
– (1983): Les types de Caméléonidés (Reptiles, Sauriens) du Muséum national d'Histoire naturelle. Catalogue critique. – Bull. Mus. Hist. nat., Paris, **5**: 1–26.
– & DOMERGUE, CH. A. (1968): Les caméléons a rostre impair et rigide de l'ouest de Madagascar. – Mém. Mus. Hist. nat., **52**: 71–110.
– & – (1969): *Chamaeleo balteatus* DUM. et BIB. (in C. et A. DUMÉRIL 1851) n'est pas synonyme de *Ch. bifidus* BRONGNIART 1800. Description du mâle. Diagnoses des espèces du groupe *parsonii*. – Bull. Mus. Nation. Hist. nat., **41**: 104–116.
– & – (1969): Notes sur les *Brookesia* de Madagascar, IV. Une série de petites *Brookesia* de Nosy Mangabé. – Bull. Mus. Hist. nat., (2) **41**: 833–841.
– & – (1975): Notes sur les *Brookesia* de Madagascar IX. Observations sur B. *tuberculata* MOCQUARD 1894, *B. ramanantsoai* sp.nov. et *B. peyrierasi* nom. nov. (Reptilia, Squamata, Chamaeleontidae). – Bull. Mus. Hist. nat., (3) **267** (1974), (Zool. 189): 1769–1782.
BUCHHOLZ, R. (1874): Bemerkungen über die im Camaroongebiet vorkommenden Arten von Chamaeleonen. – M.Ber. k. preuss. Akad. Wiss., **1874**: 77–89.
BURRAGE, B. R. (1973): Comparative ecology and behaviour of *Chamaeleo pumilus pumilus* (GMELIN) and *Ch. namaquensis* A. SMITH (Sauria: Chamaeleonidae). – Ann. S. Afr. Mus., **61**: 1–158.
CHABANAUD, P. (1923): Sur divers vertébrés à sang froid de la région Indo-Chinoise. – Bull. Mus. Hist.nat., **29**: 558–559.
CRESPO, E.G. (1972): Répteis de Portugal continental des colecções do Museu Bocage. – Arq. Mus. Bocage, **3**: 447–612.
DE WITTE, G. F. (1933): Batrachiens et reptiles recueillis par M. L. Burgeon au Ruwenzorie, au Kivu et au Tanganyka. – Rev. Zool. Bot. Afr., **24**: 111–123.
– (1965): Les caméléons de l'Afrique centrale. – Ann. Mus. roy. Afr. centr., **142**: 1–215.
DUMÉRIL, A. (1852): Familles des Caméléoniens ou Chélopodes. – Arch. Mus. Hist. nat., **6**: 257–262.
DUMÉRIL, A. M. C. & BIBRON, G. (1836): Erpétologie générale, **3**: 153–236.
DUMÉRIL, M. C. & DUMÉRIL, M. A. (1851): Catalogue méthodique de la collection des reptiles. – Gide & Baudry, Paris, I–IV, 1–224.
DUNCAN, F. H. (1937): On the dates of publication of the Society's „Proceedings", 1859–1926. – Proc. Zool. Soc. London, **107**: 71–84.
EASON, P., FERGUSON, G. W. & HEBRARD, J. (1988): Variation in *Chamaeleo jacksonii* (Sauria: Chamaeleontidae): Description of a new subspecies. – Copeia, **1988**: 580–590.
EMMRICH, D. (1991): Weitere Kenntnisse über das seltene Chamäleon *Bradypodion oxyrhinum* von den Uluguru-Bergen, Tansania. – Salamandra, **29**: 44–54.
– (1994): Herpetological results of some expeditions to the Nguru Mountains, Tanzania. – Mitt. Zool. Mus. Berlin, **70**: 281–300.
FITZSIMONS, V. F. (1935/36): Notes on the reptiles and amphibians collected and described from South Africa by ANDREW SMITH. – Ann. Transvaal Mus., **17**: 259–274.
– (1943): The lizards of South Africa. – Transvaal Mus. Mem., **1**: 151–174.
FLOWER, S. S. (1933): Notes on the recent reptiles and amphibians of Egypt, with a list of the species recorded from that kingdom. – Proc. zool. Soc., London, **1933**: 735–851.
FRITZ, J. P. & SCHÜTTE, F. (1987): Zur Biologie jemenitischer *Chamaeleo calyptratus* DUMÉRIL & DUMÉRIL 1851, mit einigen Anmerkungen zum systematischen Status. – Salamandra, **23**: 17–25.
FROST, D. R. & ETHERIDGE, R. (1989): A phylogenetic analysis and taxonomy of iguanian lizards (Reptilia: Squamata). – Misc. Publ. Univ. Kansas, **81**: 1–65.
GLAW, H. & VENCES, M. (1991): A field guide to the amphibians and reptiles of Madagascar. – Vences & Glaw-Verlag, Köln, 331 pp.
GRAVENHORST, J. L. C. (1807): Vergleichende Übersicht des Linneischen und einiger neuern zoologischen Systeme, nebst dem eingeschalteten Verzeichnisse der zoologischen Sammlung des Verfassers und den Beschreibungen neuer Thierarten, die in derselben vorhanden sind. Göttingen, Heinrich Dieterich, 476 p.
GRAY, J. E. (1831): Synopsis Reptilium. In: GRIFFITH's Animal Kingdom, **9**: 53–54, 152–155, 232–237.
– (1845): Catalog of the specimens of lizards in the British Museum.

– (1865): Revision of the genera and species of Chamaeleonidae, with the descripiton of some new species. – Proc. zool. Soc. London, **1864**: 465–479.

GROHMANN, F. S. (1832): Nuova descrizione del Camaleonte siculo. – Giorn. Sci. Lett. Arti, Palermo, **39**: 145–153.

GUIBÉ, J. (1954): Catalogue des types de lézards du Muséum national d'Histoire naturelle. – Colas (imp.), Bayeux, 119 pp.

GÜNTHER, A. (1882): Ninth contribution to the knowledge of the fauna of Madagascar. – Ann. Mag. nat. Hist., **9**: 262–266.

HELLMICH, W. (1956): Die von Dr. CHRISTA LINDEMANN und NINA PAVLITZKI in Tanganjika gesammelten Chamaeleons. – Veröff. Zool. Staats-Samml. München, **4**: 117–124.

HENKEL, F. W. & HEINECKE, S. (1993): Chamäleons im Terrarium. – Landbuch-Verlag, Hannover, 158 pp.

HEWITT, J. (1935): Some new forms of batrachians and reptiles from South Africa. – Rec. Albany Mus., **4**: 283–357.

HILLENIUS, D. (1959): The differentiation within the genus *Chamaeleo* LAURENTI 1768. – Beaufortia, Amsterdam, **8**: 1–92.

– (1963): Notes on chamaeleons II. *Chamaeleo laevigularis* (L. MÜLLER 1926), a synonym of *Chamaeleo johnstoni* (BOULENGER 1901). – Beaufortia, Amsterdam, **10**: 44–47.

– (1966): Notes on chamaeleons III. The chameleons of southern Arabia. – Beaufortia, Amsterdam, **13**: 91–108.

– (1978): Notes on chameleons V. The chameleons of North Africa and adjacent countries, *Chamaeleo chamaeleon* (LINNAEUS) (Sauria: Chamaeleonidae). – Beaufortia, Amsterdam, **28**: 37–55.

– & GASPERETTI, J. (1984): Reptiles of Saudi Arabia. The chameleons of Saudi Arabia. – Fauna of Saudi Arabia, **6**: 513–527.

HOFMAN, A., MAXSON, L. R. & ARNTZEN, J. W. (1991): Biochemical evidence pertaining to the taxonomic relationships within the family Chamaeleonidae. – Amphibia-Reptilia, **12**: 245–265.

International Code of Zoological Nomenclature (1985), International Trust for Zoological Nomenclature, London, 3rd, ed.

International Federation of Library Associations (IFLA 1977): Names of persons: national usages for entry in catalogues. 3rd ed., IFLA, London, 193 pp.

KLAVER, CH. (1973): Lung-anatomy: aid in chameleon-taxonomy. – Beaufortia, Amsterdam, **20**: 155–157.

– (1981): Lung-morphology in the Chamaeleonidae (Sauria) and its bearing upon phylogeny, systematics and zoogeography. – Z. zool. Syst. Evolut. f., **19**: 36–58.

– & BÖHME, W. (1986): Phylogeny and classification of the Chamaeleonidae (Sauria), with special reference to hemipenis morphology. – Bonn. zool. Monogr., **22**: 1–64.

– & – (1988): Systematics of *Bradypodion tenue* (MATSCHIE 1892) (Sauria: Chamaeleonidae) with a description of a new species from the Uluguru and Uzungwe Mountains, Tanzania – Bonn. zool. Beitr., **9**: 381–393.

– & – (1992): The species of the *Chamaeleo cristatus* group from Cameroon and adjacent countries, West Africa. – Bonn. zool. Beitr., **43**: 433–476

KNOEPFFLER, L. PH. (1967): Redécouverte de *Chamaeleo chapini* DE WITTE. – Biol. Gabonica, **3**: 245–247.

KUHL, H. (1820): Beiträge zur Kenntnis der Amphibien. In: Beiträge zur Zoologie und vergleichenden Anatomie. – Ed.: H. KUHL, Frankfurt a. M.: 77–132.

LOVERIDGE, A. (1929): East African reptiles and amphibians in the United States National Museum. – U.S. Nation. Mus. Bull., **151**: 1–135.

– (1933a): Reports on the scientific results of an expedition to the southwestern highlands of Tanganyika Territory. VII. Reptiles. – Bull. Mus. comp. Zool., **74**: 197–416.

– (1933b): Reports on the scientific results of an expedition to the southeastern highlands of Tanganyika Territory. I. Introduction and zoogeography. – Bull. Mus. comp. Zool., **75**: 3–43.

– (1942): Scientific results of a fourth expedition to forested areas in East and Central Africa. – Bull. Mus. comp. Zool., **91**: 237–373.

– (1944): Scientific results of a fourth expediton to forested areas in East and Central Africa. VI. Itinerary and comments. – Bull. Mus. comp. Zool., **94**: 191–214.

– (1953): Zoological results of a fifth expediton to East Africa. III. Reptiles from Nyasaland and Tete. – Bull. Mus. comp. Zool., **110**: 143–322.
– (1956): A new subgenus of *Chamaeleo* from Rhodesia and new races of *Mabuya* from Kenya Colony. – Breviora, **59**: 1–4.
– (1957a): Checklist of the reptiles and amphibians of East Africa (Uganda; Kenya; Tanganyika; Zanzibar). – Bull. Mus. comp. Zool., **117**: 153–362.
– (1957b): On a third collection of reptiles taken in Tanganyika by C. J. P. IONIDES, Esq. – Tanganyika Notes & Records, **43**: 1–19.
– (1959): On a fourth collection of reptiles, mostly taken in Tanganyika Territory by Mr. C. J. P. IONIDES. – Proc. zool. Soc. London, **133**: 29–44.
MCKEOWN, S. (1991): Jackson's chameleons in Hawaii are the recently described Mt. Kenya subspecies. – Bull. Chicago Herpetol. Soc., **26**: 49.
MEERMAN, J. & BOOMSMA, T. (1987): Beobachtungen an *Chamaeleo calyptratus calyptratus* DUMÉRIL & DUMÉRIL 1851 in der Arabischen Republik Jemen. – Salamandra, **23**: 10–16.
MERTENS, R. (1922): Verzeichnis der Typen in der herpetologischen Sammlung des Senckenbergischen Museums. – Senckenbergiana, **4**: 162–183.
– (1938): Herpetologische Ergebnisse einer Reise nach Kamerun. – Abh. senckenb. naturf. Ges., **442**: 1–52.
– (1940): Zur Herpetologie Kameruns und Deutsch-Ostafrikas. – Zool. Anz., **131**: 239–250.
– (1949): Lebendgebärende Dreihorn-Chamäleons. – Natur u. Volk, **79**: 195–202.
– (1955): Amphibien und Reptilien aus Ostafrika. – Jh. Ver. vaterl. Naturk. Württ., **110**: 47–61.
– (1959): Über das Vierhorn-Chamäleon, *Chamaeleo quadricornis*. – Natur u. Volk, **89**: 114–118.
– (1966): Chamaeleonidae. – Das Tierreich, **83**: I–X, 1–37.
– (1967): Die herpetologische Sektion des Natur-Museums und Forschungs-Institutes Senckenberg in Frankfurt a. M., nebst einem Verzeichnis ihrer Typen. – Senckenbergiana, **48**: 1–106.
– (1969): Die Amphibien und Reptilien West-Pakistans. – Stuttgarter Beitr. Naturk., **197**: 1–96.
– & MÜLLER, L. (1928): Liste der Amphibien and Reptilien Europas. – Abh. senckenb. naturf. Ges., **41**: 1–62.
– & WERMUTH, H. (1960): Die Amphibien und Reptilien Europas. Dritte Liste, nach dem Stand vom 1. Januar 1960. – Kramer, Frankfurt a. M., 264 pp.
METHUEN, P. A. & HEWITT, J. (1914): Contributions to our knowledge of the anatomy of chamaeleons. – Trans. roy. Soc. S. Afr., **4**: 89–104.
MOCQUARD, F. (1894): Diagnoses de quelques reptiles nouveaux de Madagascar. – C. R. Soc. philom., **9**: 3–10.
– (1909): Synopsis du familles, genres, espèces des reptiles écailleux et batrachiens de Madagascar. – Nouv. Arch. Mus. Hist. nat., **1**: 1–100.
NEČAS, P. (1994): Bemerkungen zur Chamäleon-Sammlung des Naturhistorischen Museums in Wien, mit vorläufiger Beschreibung eines neuen Chamäleons aus Kenia. – Herpetozoa, **7**: 95–108.
PERRET, J. L. & MERTENS, R. (1957): Etude d'une collection herpétologique faite au Cameroun de 1952 à 1955. – Bull. Inst. franç. Afr. noire, Dakar, **19**: 548–601.
PARKER, H. W.; MOREAU, R. E. & PAKENHAM, P. H. W. (1940): Notes on the herpetology of the East African Islands. – Ann. Mag. nat. Hist., **5**: 309–314.
POWER, J. H. (1932): On the South African species of chameleons of the genus *Lophosaura*. – Proc. zool. Soc. London **1932**: 209–224.
RAMANANTSOA, G. (1974): Connaissance des caméléonidés commun de la province de Diégo Suarez par la population paysanne. – Bull. Acad. malgache, **51**: 147–149.
RAND, A. S. (1963): Notes on the *Chamaeleo bitaeniatus* complex. – Bull. Mus. comp. Zool., **130**: 1–29.
RAW, L. R. G. (1976): A survey of the dwarf chameleons of Natal, South Africa, with descriptions of three new species. – Durban Mus. Novit., **11**: 139–161.
– (1978): A further new dwarf chameleon from Natal, South Africa. – Durban Mus. Novit., **11**: 265–269.
RAXWORTHY, C. J. (1991): Field observations on some dwarf chameleons (*Brookesia* spp.) from

rainforest areas of Madagascar, with the description of a new species. – J. Zool., **224**: 11–25.

– & NUSSBAUM, R. A. (1995): Systematics, speciation and biogeography of the dwarf chameleons (*Brookesia*; Reptilia, Squamata, Chamaeleontidae) of northern Madagascar. – J. Zool. **235**: 525–558.

RIEPPEL, O. (1987): The phylogenetic relationships within the Chamaeleonidae, with comments on some aspects of cladistic analysis. – Zool. J. Linn. Soc., **89**: 41–62.

SALDANHA, L. (1978): Museu Bocage. – Copeia, **1978** (4): 739–740.

SEBA, A. (1734): Locupletissimi rerum naturalium thesauri accurata descriptio, et iconibus artificiosissimis expressio, per universam physices historiam. **I**: 1–178.

SCHMIDT, K. P. (1919): Contributions to the herpetology of the Belgian Congo, based on the collection of the American Congo Expedition, 1909–1915. – Bull. Amer. Mus. nat. Hist., **39**: 384–624.

STEJNEGER, L. (1936): Types of the amphibian and reptilian genera proposed by LAURENTI in 1768. – Copeia, **1936**: 131–141.

STERNFELD, R. (1912): Reptilia. – Wiss. Ergebn. dtsch. Zentral-Afrika-Exped. 1907–1908. – Ed.: SCHUBOTZ, H., IV: 197–279.

STEVENS, R. A. (1974): An annotated check list of the amphibians and reptiles known to occur in south-eastern Malawi. – Arnoldia, **6**: 1–22.

TILBURY, C., R. (1991): A new species of *Chamaeleo* LAURENTI 1768 (Reptilia Chamaeleonidae) from a relict montane forest in northern Kenya. – Trop. Zool., **4**: 159–165.

– (1992): A new dwarf forest chameleon (Sauria: *Rhampholeon* GÜNTHER 1874) from Malawi, central Africa. – Trop. Zool., **5**: 1–9.

TORNIER, G. (1889): Neues über Chamaeleons. – Zool. Anz., **22**: 408–414.

– (1902): Herpetologisch Neues aus Deutsch-Ost-Afrika. – Zool. Jb., Syst., **15**: 578–590.

WILD, C. (1993): Notes on the rediscovery and congeneric associations of the PFEFFER's chameleon *Chamaeleo pfefferi* (TORNIER 1900) (Sauria: Chamaeleonidae), with a brief description of the hitherto unknown female of the species. – Brit. Herpetol. Soc. Bull., **45**: 25–32.

WERNER, F. (1902): Prodromus einer Monographie der Chamäleonten. – Zool. Jb., Syst., **15**: 295–460.

– (1911): Chamaeleontidae. – Das Tierreich, **27**: I–XI, 1–52.

Alphabetical index

As of Part 102 **Das Tierreich** is published in hard cover. Since 1958 the following have appeared

Contribution	Author	Title	Year	No. of pages
72	H. Wermuth	REPTILIA: *Helodermatidae*	1958	16 pages, 9 figures
73	M. Beier	ORTHOPTERA: *Tettigoniidae* (Pseudophyllinae I)	1962	XII, 468 pages, 245 figures
74	M. Beier	ORTHOPTERA: *Tettignoniidae* (Pseudophyllinae II)	1960	396 pages, 241 figures
75	K. Odening	TREMATODA (Digenea): *Plagiorchiidae* III (Haematoloechinae) und *Omphalometridae*	1960	IV, 77 pages 60 figures
76	K. Sanft	AVES (Upupae): *Bucerotidae*	1960	IV, 176 pages 106 figures
77	R. Kilias	GASTROPODA PROSOBRANCHIA: *Tonnidae*	1962	63 pages, 47 figures
78	St. W. Gorham	GYMNOPHIONA, *Liste der rezenten Amphibien und Reptilien*	1962	X, 25 pages
79	R. Mertens	HELODERMATIDAE, VARANIDAE, LANTHANOTIDAE. *Liste der rezenten Amphibien und Reptilien*	1963	X, 26 pages
80	H. Wermuth	GEKKONIDAE, PYGOPODIAE, XANTUSIIDAE. *Liste der rezenten Amphibien und Reptilien*	1965	XXII, 246 pages
81	J. A. Peters	COLUBRIDAE (Dipsadinae). *Liste der rezenten Amphibien und Reptilien*	1965	VIII, 18 pages
82	J. Illies	PLECOPTERA, *Katalog*	1966	XXX, 632 pages, 20 figures
83	R. Mertens	CHAMAELEONIDAE. *Liste der rezenten Amphibien und Reptilien*	1966	X, 37 pages
84	C. Gans	UROPELTIDAE. *Liste der rezenten Amphibien und Reptilien*	1966	29 pages
85	St. W. Gorham	ASCAPHIDAE, LEIOPELMATIDAE, PIPIDAE, DISCOGLOSSIDAE, PELOBATIDAE, LEPTODACTYLIDAE, RHINOPHRYNIDAE. *Liste der rezenten Amphibien und Reptilien*	1966	XVI, 222 pages
86	H. Wermuth	AGAMIDAE. *Liste der rezenten Amphibien und Reptilien*	1967	XIV, 127 pages
87	H. Wermuth	CORDYLIDAE (Cordylinae + Gerrhosaurinae). *Liste der rezenten Amphibien und Reptilien*	1968	X, 30 pages
88	F. Haas	BIVALVIA: *Unionacea*	1969	X, 663 pages, 5 figures

Contribution	Author	Title	Year	No. of pages
89	A. F. Stimson	BOIDAE, (Boinae + Bolyeriinae + Loxoceminae + Pythoninae). *Liste der rezenten Amphibien und Reptilien*	1969	XI, 49 pages
90	H. Wermuth	ANGUIDAE, ANNIELLIDAE, XENOSAURIDAE. *Liste der rezenten Amphibien und Reptilien*	1969	XII, 41 pages
91	M. Fischer	HYMENOPTERA: *Braconidae* (Opiinae I)	1972	XII, 620 pages, 463 figures
92	R. Kilias	GASTROPODA PROSOBRANCHIA: *Cymatiidae*	1973	VIII, 235 pages, 149 figures
93	P. Banarescu & T. T. Nalbant	PISCES, TELEOSTEI: *Cyprinidae (Gobioninae)*	1973	VII, 304 pages, 154 figures, 19 maps
94	P. Zwick	PLECOPTERA, *Phylogenetisches System und Katalog*	1973	XXXII, 465 pages, 75 figures
95	E. Duellmann	HYLIDAE, CENTROLENIDAE, PSEUDIDAE. *Liste der rezenten Amphibien und Reptilien*	1976	XIX, 225 pages
96	M. Fischer	HYMENOPTERA: *Braconidae* (Optiinae II/Amerika)	1976	1038 pages, 890 figures
97	S. G. Kiriakoff	LEPIDOPTERA NOCTUIFORMES: *Agaristidae I*	1976	IX, 180 pages, 39 figures
98	S. G. Kiriakoff	LEPIDOPTERA NOCTUIFORMES: *Agaristidae II*	1976	VIII, 165 pages, 52 figures
99	S. G. Kiriakoff	LEPIDOPTERA NOCTUIFORMES: *Agaristidae III*	1976	VI, 86 pages 26 figures
100	H. Wermuth & † R. Mertens	TESTUDINES, CROCODYLIA, RHYNCHOCEPHALIA. *Liste der rezenten Amphibien und Reptilien*	1976	XXVII, 174 pages
101	D. E. Hahn	ANOMALEPIDIDAE, LEPTOTYPHLOPIDAE, TYPHLOPIDAE. *Liste der rezenten Amphibien und Reptilien*	1980	XII, 93 pages
102	H. Steinmann	DERMAPTERA *Catadermaptera I*	1986	XIV, 343 pages
103	A. Kaltenbach	SAGINAE *Saltatoria-Tettigoniidae*	1986	VIII, 92 pages
104	M. Fischer	HYMENOPTERA *Optiinae III*	1987	XV, 734 pages
105	H. Steinmann	DERMAPTERA *Catadermaptera II*	1989	XIX, 504 pages

Contribution	Author	Title	Year	No. of pages
106	H. Steinmann	DERMAPTERA *Eudermaptera I*	1989	XX, 558 pages
107	R.L. Hoffman	MYRIAPODA 4 *Polydesmida: Oxydesmidae*	1990	XVI, 512 pages
108	H. Steinmann	DERMAPTERA *Eudermaptera II*	1993	XXII, 711 pages
109	A.M. Bauer	GEKKONIDAE *(Reptilia Sauria)*	1994	XIV, 306 pages
110	H. Steinmann	World Catalogue of Odonata Volume I *Zygoptera*	1997	XXII, 500 pages
111	H. Steinmann	World Catalogue of Odonata Volume II *Anisoptera*	1997	XIV, 636 pages
112	C.J.J. Klaver & W. Böhme	CHAMAELEONIDAE	1997	XVI, 85 pages

Bei Fragen zur Produktsicherheit wenden Sie sich bitte an:
If you have any questions regarding product safety,
please contact:

Walter de Gruyter GmbH
Genthiner Straße 13
10785 Berlin
productsafety@degruyterbrill.com